GARDNER'S
Whys & Wherefores

GARDNER'S
Whys & Wherefores

MARTIN GARDNER

THE UNIVERSITY OF CHICAGO PRESS
Chicago and London

Martin Gardner wrote the Mathematical Games column in *Scientific American* for twenty-five years. He is the author of many books on science, mathematics, magic, philosophy, and literary criticism, including *The Annotated Alice* and *The Whys of a Philosophical Scrivener*. His books published by the University of Chicago Press include *The Annotated "Casey at the Bat," Logic Machines and Diagrams, Hexaflagons and Other Mathematical Diversions, The Second* Scientific American *Book of Mathematical Puzzles and Diversions, Martin Gardner's New Mathematical Diversions,* and *Martin Gardner's Sixth Book of Mathematical Diversions.*

The University of Chicago Press, Chicago 60637
The University of Chicago Press, Ltd., London
© 1989 by The University of Chicago
All rights reserved. Published 1989
Printed in the United States of America

98 97 96 95 94 93 92 91 90 89   5 4 3 2 1

Library of Congress Cataloging-in-Publication Data

Gardner, Martin, 1914–
    Gardner's whys & wherefores.

    I. Title. II. Title: Gardner's whys and wherefores.
PS3557.A714G37   1989   814'.54   88–20624
ISBN 0–226–28245–7

∞ The paper used in this publication meets the minimum requirements of the American National Standard for Information Sciences—Permanence of Paper for Printed Library Materials, ANSI Z39.48-1984.

In Memoriam: Bob and Betty

Contents

Preface

Order and Surprise (Prometheus Books, 1983) was a selection of essays and book reviews written over a period of forty years. Aside from the first four chapters, this is a similar collection of pieces written after the publication of that book. Part I reprints essays; Part II reprints reviews. Within each part, the chapters are in chronological order of publication.

Martin Gardner

PART ONE
ESSAYS

The Ancient Mariner

But I do not think "The Rime of the Ancient Mariner" was for Coleridge an escape from reality: I think it was reality, I think he was on
the ship and made the voyage and felt and knew it all.

> —THOMAS WOLFE, in a letter of 1932, included in *The Letters of
> Thomas Wolfe,* edited by Elizabeth Nowell (Charles Scribner's Sons, 1956,
> p. 322)

"A poem," so runs a much quoted line by Archibald MacLeish,
"should not mean but be." It is a puzzling statement. How can a
poem, unless it means something, possibly "be"? Other types of art
are quite different. A symphony does not have to mean anything;
the listener has a direct, pleasurable experience of the sounds. An
abstract painting does not have to mean anything; it is just there, a
created, palpable object to be observed and enjoyed. But a poem
has to be communicated by queer little black marks on white paper,
the marks arranged in complicated patterns that give no aesthetic
pleasure in themselves. The patterns are visual symbols understood
only by a person with a sensitive memory of how his culture attributes sounds and meanings to those patterns. Without these sounds
and meanings, there is no poem.

Nevertheless, MacLeish's line does make a significant point.
The little black marks are not the poem. They are no more than
patterns employed to symbolize the poem. After one has interpreted these patterns as best he can, drawing on all the subtle
sounds and meanings that a culture has bestowed on them, the
poem itself—the real poem—takes shape as a constructed object,
a thing (even though it exists only in his mind) that can be directly
experienced in a way not much different from the way in which one
experiences a symphony or a painting. When that stage is reached,
the poem ceases to be something that must be explained; it becomes an art object to be experienced.

The purpose of the notes in my *Annotated Ancient Mariner* is
to help the reader reach such a stage with respect to one poem. My
attempts to explicate Coleridge's ballad were relatively superficial:
defining words and phrases, clarifying obscurities that have arisen

This article first appeared as the afterword of my now long out of print *Annotated
Ancient Mariner* (Clarkson Potter, 1965).

because we are not living in early nineteenth-century England, and pointing out subtleties of meaning probably intended by the poet but easily missed unless the poem is read many times and thought about deeply. The notes also touched occasionally on the techniques by which Coleridge strengthened the vividness and emotional impact of his lines. We saw how he chose the old English ballad form to convey the feel of an ancient time; how he borrowed from this form its use of repetition, alliteration, elemental imagery, color, and archaic diction. We saw how he avoided the rigidity of the old ballad form by skillfully varying its rhythms and the number of lines in its stanzas, whenever such departures contributed to a desired effect. Above all, we saw how Coleridge, with his great sensitivity to the rich overtones of English, chose his words and phrases, and put them together in such a way, as to arouse intensely vivid pictures in the mind. It is this extraordinary power of Coleridge that provides the sheer "magic" of the poem. (No other adjective has been more often applied to it.) Disbelief in the Mariner's preposterous tale is momentarily put aside; one almost *sees* the fire-flags and the water snakes, the helmsman's face illuminated by the lamp, the motionless weathercock, the bloodred sun, the stony, glittering eyes of the dead men. No English poem before or since has been capable of arousing, for so many readers, such intense images of unearthly beauty and terror.

Moreover, there is a curious way in which the imagery of a great poem such as this grows even more intense with the passage of time. Its lines and episodes work their way into the literature of a culture: they are quoted, borrowed, and echoed by later poets and writers. There is a kind of feedback. When we read a poem that has become a classic, its lines reverberate with subliminally comprehended overtones that derive from later works of literature. Every great poem suffers an inevitable erosion of meaning with the passage of years, as language and customs and values change; but, at the same time, every great poem accumulates new meanings. In some respects *The Ancient Mariner* can be read today with greater pleasure than it could in Coleridge's time.

A few notes have pointed out ways in which events in Coleridge's life and aspects of his personality may have colored the meanings of certain lines. Though such biographical analysis is often considered superficial, it does add something, however small, to the total meaning of a poem. Hugh I'Anson Fausset's *Samuel Taylor Coleridge* (1926) devotes a chapter to *The Ancient Mariner* in which this approach is stressed. When the Mariner speaks of passing like night from land to land, with his strange power of speech,

Fausset sees Coleridge himself, "longing to escape from the solitude of an abnormal consciousness, seeking relief throughout his life in endless monologues." When the Mariner speaks of walking to the kirk as sweeter than the marriage feast, is Coleridge thinking of "his own never-satisfied need of simple, devout human relationships . . ."? Is the stanza that begins "Oh sleep! it is a gentle thing" an expression of the poet's physical indolence? Is the moral ("He prayeth best . . .") an expression of "his own childlike affection for everything without distinction"? And so on. Fausset is convinced that it is because Coleridge projected his own hopes and terrors so completely into this ballad that it acquired more "reality" than his other poems and became his greatest poem.

Lowes's approach to the ballad, in *The Road to Xanadu* (1927), is also essentially biographical. As the reader surely knows, this famous tour-de-force of English criticism is an exhaustive study of the ballad's literary sources. Following up clues in Coleridge's notebooks and letters, Lowes set himself the task of trying to read everything that Coleridge was known to have read before he wrote the poem, as well as books he probably read or even *might* have read. This literary detective work paid off handsomely, for it turned out that Coleridge had borrowed heavily, often exact words and phrases, from the leading sea travel books of the time. The more important of these borrowings have been cited in our notes.

In his omnivorous reading of travel books, Coleridge always read with what Lowes called the poet's "falcon eye," searching for just those details which could best be transmuted into poetry. But Lowes believed that these details worked their way into Coleridge's unconscious and preconscious, where they lay dormant until he began the actual writing of his ballad. Then, by those obscure processes of association which had been detailed and analyzed by David Hartley, memories of what he read fused together in his conscious mind and emerged in lines of the ballad. Maybe so. I am inclined, however, to suspect that it was all more conscious than Lowes would have us believe. There is no reason why Coleridge, when he reached the point at which he wanted to describe the colors of icebergs, could not have flipped through the pages of a travel book until he found, say, the phrase "green as emerald." "Perfect!" he shouts. "Just what I've been looking for!"

This is the point of view taken by Robert Cecil Bald in his important paper, "The Ancient Mariner: Addenda to *The Road to Xanadu*," in *Nineteenth Century Studies* (1940), Coleridge's reading, says Bald (drawing on data from notebooks not available to Lowes), was not as random as Lowes proposes. One of Coleridge's notes is: "To read most carefully for the purposes of poetry" an account of an earthquake. Coleridge may have been fully aware of how much he had borrowed from the travel books.

Evaluating the influence of opium on Coleridge's poems is also part of a biographical approach. Here, too, Bald takes issue with Lowes. Lowes played down the effects of the drug. John M. Robertson, in *New Essays Toward a Critical Method* (1897), and Meyer H. Abrams, in *The Milk of Paradise* (1934), play up the drug. Bald steers a middle course. He reasons, quite sensibly, that although Coleridge may not have been strongly addicted to opium at the time he wrote *The Ancient Mariner,* he may have been sufficiently addicted to experience the milder reveries. Perhaps in 1800, when he added the subtitle "A Poet's Reverie" to the ballad, he was thinking of opium reveries. Perhaps one reason he seized on the plot, when it was suggested by a neighbor's dream, was because he saw at once that here was a magnificent outlet for the vivid dream phenomena that had been building up in his mind. There is no way to be sure, but this view seems plausible.

Another important biographical fact, already mentioned in the notes, is that Coleridge had never been to sea when he first wrote his ballad. During World War II, I served in the North Atlantic on a destroyer escort, a ship small enough so that a sailor could really get to know the sea in a way quite different from that of the tourist who floats gently over the ocean on a huge hotel. I can assure the reader that the smell of the sea is not in the first version of Coleridge's poem. This is not to say that the poem does not convey a strong sense of reality but only to say that it is not the real sea that Coleridge makes seem real. It is a fantasy sea. There is less of the true sea in the entire ballad than in a dozen lines that one can find on hundreds of pages by Melville or Conrad, or in Masefield's single line: "And the flung spray and the blown spume and the seagulls crying." "This great sea-piece might have had more in it of the air and savour of the sea," wrote Swinburne (discussing the poem in *Memories and Studies*). David Hartley's doctrine of association of ideas explains, of course, why the sea's savor never entered the poem's first version: Coleridge had never savored it.

As Bald points out, Coleridge later *did* go to sea, and then he did add to his ballad some passages which more strongly convey a sense of actual sea life. For example, the lines about the sloping masts of the ship as the storm blast drives it forward and the description of the helmsman's face lit at night by his lamp. (There is a notebook entry, No. 2001, in which Coleridge records having observed the second scene.) But even with these additions, the ship remains largely a painted ship on a painted ocean.

An attempt to probe more deeply into the influence of Coleridge's life and character on the poem has been made by David Beres in his paper, "A Dream, a Vision, and a Poem" (*International Journal of Psycho-Analysis,* Vol. 32, Part 2 [1951]). As far as I know, it remains the only attempt by a professional analyst to explicate the poem by way of Freudian insights and symbols. Coleridge's many references in his writings to food and hunger, his strong desire to be loved, his preoccupation with sleep and dreams, his "devouring" of books, and so on, lead Beres to conclude that Coleridge was an almost clinically perfect instance of what the Freudians call an oral character. In addition, he finds evidence that early in life Coleridge developed an ambivalent attitude toward his mother, failed to resolve his infantile aggression, became confused as to his sexual identity. He sees Coleridge's relationships with his male friends, especially Wordsworth, as unconsciously homosexual.[1] The poet's inability to resolve his sense of guilt toward his mother thus underlies his unhappy love life, his steadily increasing depression and anxiety, and his dependence on opium as a relief from suffering.

How accurate this picture is I am not prepared to say. Beres advanced it tentatively, before he had access to Coleridge's notebooks, now being published. In many ways the picture seems accurate, though a careful student of Coleridge is likely to suspect that the poet's character is less simple than Beres makes it out to be. I will cite only one instance of how easy it is to misinterpret biographical details. Coleridge's poem "Dejection: An Ode" is a moving expression of grief, despair, and a sense of one's creative powers slowly being drained away. "It cannot be without significance," writes Beres, "that this poem was written on the occasion of Wordsworth's marriage to Mary Hutchinson, an added hint of Coleridge's unconscious homosexual attachment to his friend." But the poem was *not* written on the occasion of Wordsworth's marriage; it just happened to be *published* in a newspaper on the day of the marriage. It was written six months before Wordsworth's mar-

---

1. "I believe it possible that a man may, under certain states of the moral feeling, entertain something deserving the name of love towards a male object—an affection beyond friendship, and wholly aloof from appetite." Havelock Ellis quotes this passage (from Coleridge's *Table Talk* [May 14, 1833]) in his book on *Sexual Inversion* and comments: "This passage of Coleridge's is interesting as an early English recognition by a distinguished man of genius of what may be termed ideal homosexuality."

riage and has long been known to be an expression of Coleridge's hopeless, unrequited love for Sarah Hutchinson. The published version was carefully cut and edited to conceal the true source of his grief, but the original version of the poem, which he sent in a letter to Sarah, has been known since 1937, when it was first printed by Ernest de Selincourt. (Selincourt later included it in his *Wordsworthian and Other Studies,* 1947. It may also be found in George Whalley's *Coleridge and Sara Hutchinson,* 1955, and Humphry House's *Coleridge,* 1953.) One could scarcely find a more "clinically perfect" description of that drained, empty feeling of a man who has come to realize that a love for a certain woman is not, and will not, be returned. It was not Coleridge but Dorothy Wordsworth who was most upset by her brother's marriage.

It is harder to take seriously Beres's symbolic interpretation of *The Ancient Mariner.* The albatross naturally is regarded as a symbol of Coleridge's mother, providing the poet an outlet for his repressed hostility. The Mariner's crime is mother-murder, carrying with it, according to Beres, an unconscious incest motive. (For Kenneth Burke, in *Philosophy of Literary Form,* the bird is a symbol of Coleridge's wife. Beres would probably agree, for he regards the wife herself as another mother symbol.) His punishment is hunger, loss of love, and loneliness: "pregenital punishments for a preoedipal crime." The mother image appears again in the ballad as the avenging spectre-woman, Life-in-Death, then finally as the forgiving Holy Mother who sends rain and sleep. "The mother whom he restores to life brings him back to the safety of his homeland. A mother-figure forgives the crime against the mother."

With this background, Beres has little difficulty interpreting numerous lines in ways that reinforce his central theme. Consider the "silly buckets on the deck" (line 297). Beres points out that according to the *Shorter Oxford Dictionary* the meaning of "silly" is "feeble, frail, insignificant." He concludes: "It is not too rash an assumption that the buckets symbolize the mother's breasts, previously empty and cruel, now full and forgiving."

Well, perhaps a *bit* rash? This is not meant to question the soundness of Beres's over-all analysis of Coleridge's character but to question the degree to which unconscious symbolic meanings can be made specific in a fantasy poem so rich in symbolic possibilities. Rules for the interpretation of literary symbols, in the Freudian school, are so loose and uncontrolled that attempts at such interpretation easily degenerate into a clever game based largely on word puns and visual similarities. The game is so easy to play that it becomes almost valueless in providing reinforcement for a character interpretation.

Let the reader try the following experiment. Pick at random any type of neurotic personality. Then go over *The Ancient Mariner* carefully, searching for symbols to reinforce the traits of the chosen personality. The reader will be amazed at how readily the symbols turn up. The point is not that a poem cannot contain unconscious symbolic expressions of the poet's hopes, fears, and conflicts but that it is rash to regard the finding of such symbols as confirming evidence for previously made character assumptions. Analysts who play this game with literature seem to have little awareness of how fantastically flexible are the controls on this kind of feedback—so flexible, in fact, that it is impossible to lose.

For example, if Coleridge had spoken of the albatross as "her," Beres would have considered this supporting evidence for his central thesis. But Coleridge speaks of the bird as "him" (line 405). No matter—one wins either way. "To Coleridge," writes Beres, in accounting for the "him," "the father was a feminine, giving male; the mother a masculine, rejecting female." Had Coleridge not specified the bird's sex, this too would do the trick, for was not the mother an ambivalent sexual figure? The Freudian critic first makes a tentative hypothesis about a poet's neuroses, he searches for symbols in the poet's writings that fit the hypothesis, then the symbols are taken as confirming evidence for the hypothesis. I do not here attack the value of the hypothesis; I merely suggest (in a whisper) that Freudian symbolic interpretation is so elastic and uncontrolled that the support it provides for character analysis is largely a mirage. This is especially true with respect to a fantasy poem in which there are hundreds of possible symbols, each capable of many interpretations.

The analytic tradition, on its Jungian side, provides the background for Maud Bodkin's symbolic approach to *The Ancient Mariner* in her path-breaking study, *Archetypal Patterns in Poetry* (1934). Miss Bodkin is less concerned with the poem as an expression of Coleridge's neuroses than with the poem as an expression of certain universal emotional patterns that are inescapable aspects of human nature itself, as distinct from less basic emotions that may vary from time to time, culture to culture, person to person. She reasons that, when a poem continues to fascinate and deeply move large numbers of readers for more than a century, it must deal with emotional themes of great universality—themes that are permanently impressed upon the reader's unconscious. From Jung she borrows the term "archetype" as a name for such a theme, and with this she also takes over, though with certain reservations and doubts, Jung's belief in a "collective unconscious."

Both Freud and Jung were firmly convinced that the human mind, at birth, has stamped upon its neural circuits various patterns of behavior and emotional response that are records of the collective experience of countless ancestors. This aspect of the analytic movement has today been abandoned (except for a few old-fashioned fundamentalists), but in 1934, when Miss Bodkin published her book, the anthropological evidence was only starting to pour in, and the revisionary work of the more progressive analysts was only beginning. It is to Miss Bodkin's credit that she was well enough acquainted with the anthropological evidence to admit that there was no empirical support for Jung's collective unconscious and that it might be possible to explain the universality and persistence of archetypal patterns in terms of a common cultural heritage and common aspects of human experience. Of course it is possible. The entire biological background of Miss Bodkin's book can be discarded with no loss whatever to her critical positions.

Consider, for example, the archetype that she finds central to the emotional power of *The Ancient Mariner:* the theme of death and rebirth. The Mariner commits a senseless crime, a "hellish thing." It results in the death of all his shipmates and plunges him into physical suffering and mental agony. He finds himself alone on a rotting sea, dying of thirst, surrounded by a thousand thousand slimy things. He tries to pray, but his heart is as dry as his throat. "The imagery of calm and drought here," writes R. L. Brett in *Reason and Imagination* (1960), "is as old as religious poetry itself. From the valley of dry bones in the Book of Ezekiel to Eliot's *The Waste Land,* dryness has symbolized spiritual barrenness."

Then comes the turning point, that great somersault of faith so cleverly symbolized by Dante when, at the exact center of the earth, he turns himself upside down and begins the slow climb from hell to purgatory. The Mariner's heart goes out in love toward God and nature. He is suddenly able to pray. The albatross, symbol of his burden of guilt, drops from his neck. The gentle peace of heaven slides into his soul, and when he awakes, rain assuages his great thirst. The ship starts its mysterious motion that carries him home:

> But soon there breathed a wind on me,
> Nor sound nor motion made:
> Its path was not upon the sea,
> In ripple or in shade.

> It raised my hair, it fanned my check
> Like a meadow-gale of spring—
> It mingled strangely with my fears,
> Yet it felt like a welcoming.

The poem is thus an allegory in the higher sense in which certain Greek myths may be considered allegorical. Coleridge himself, in his criticism, stressed the difference between what he called an allegory—a story in which every object and incident is symbolic of something else—and a symbolic narrative in which only certain objects and incidents give to the story, in an overall way, a universal meaning. The ballad, in short, is a myth. It is more like, say, *Moby Dick* than *Pilgrim's Progress* or *The Faerie Queene*. It can be read, understood, and enjoyed solely as a narrative, but after many readings and much reflection, a higher level of significance forms above the narrative like a luminous cloud.

There is little doubt that a major source of the ballad's emotional power is this rebirth pattern of sin-suffering-death-repentance-rebirth-penance-salvation. But there is no need to invoke the Lamarckian views of nineteenth-century biologists to explain the power of such a pattern. Not only is it one of the central myths of Christianity, the shreds of which are still firmly a part of the West's cultural heritage, but the bare bones of the rebirth theme are inescapable in every person's experience. Each night we lapse into unconsciousness, a death of many hours, to find ourselves reborn the following morning. Every year we see the wintry deaths of trees and plants followed by their spring rebirths. We take to bed with an illness, we recover. Older generations pass away, the young take over. Life is filled with cycles of death and renewal. There is no need to invent a process by which collective memories of the dead are recorded inside our skulls. The experience of every child before the age of ten, in Manhattan or Africa, is sufficient to account for the mind's response to the rebirth archetype in literature.

A child who grows up with a Protestant or Catholic background learns, of course, versions of the rebirth pattern that have the more specific form which Coleridge's ballad dramatizes. This is true even in liberal Protestant churches where the minister and most of his congregation do not believe in the Resurrection of Christ or that conversion is a transfer from a road leading to eternal damnation to a road leading to eternal happiness. On Easter Sunday even a Unitarian minister, a bit embarrassed by the larger church attendance, inevitably finds himself speaking in symbols of death and rebirth. A child who attends a more traditional Protestant church will sing hymns, some of them dating back to Coleridge's time, which exploit symbols virtually identical with some of the rebirth symbols in *The Ancient Mariner.* The "going home" theme is central to hundreds of hymns. Some even identify the lost soul with a mariner lost at sea:

> My soul in sad exile was out on life's sea,
>   So burdened with sin and distressed,
> Till I heard a sweet voice saying, "Make me your choice,"
> And I entered the haven of rest.

Another source of the ballad's emotional power is, one suspects, its exploitation of an archetype even older and more pervasive than the rebirth pattern. I refer to the concept of the supernatural, the myth of Plato's cave, the conviction that the world we know is a shadow world. Behind it, hidden from us, is another, wholly other world. Coleridge's quotation from Burnet, which heads his poem, speaks of exactly this. The ballad is a Platonic poem, suffused throughout with that *mysterium tremendum* that lies at the heart of all the world's great religious faiths. The poem's horrors are the horrors of this world. Is the silent, rotting sea more ghastly than the steady-state universe of a modern naturalist, endlessly repeating more of the same, in all directions, throughout eternity, like one of those mad, meaningless machines that mechanics sometimes build as a joke? It is this transcendent blankness, this absolute nothingness surrounding the universe, that is the ultimate repressed horror of modern Aristotelianism. It is the Platonic dualism, the intimation of a higher reality—albeit one filled with angels and water daemons—that is the huge and omnipresent archetypal pattern of Coleridge's ballad.

Throughout the poem Coleridge plays with this and other archetypal themes and with words and images that in themselves, like the words of the King James Bible or a good folk song, are archetypal and eternal: wind, rain, sun, moon, star, sleep, soul, love, life, death. Miss Bodkin's book devotes several memorable pages to analyzing the emotional effect of "red" in that wonderful stanza:

> Her beams bemocked the sultry main,
>   Like April hoar-frost spread;
> But where the ship's huge shadow lay,
> The charmèd water burnt alway
>   A still and awful red.

"There is, I suspect," wrote Lowes, "no magic in the poem more potent than this blending of images through which the glowing redness of animalcules once seen in the Pacific has imbued with sombre mystery that still and boding sea and the image which lies across it with utter distinctness in a hush of brooding light."

Why does this simple three-letter word "red" hit the reader with such force? Because, suggests Miss Bodkin, "the word 'red' has a soul of terror that has come to it through the history of the race." Yes, but the Jungian racial memory is wholly unnecessary to explain it. Before a child talks he has associated red with the color of blood and fire; before he is a young man he has associated it with a girl's lips, the signal light of danger, the scarlet woman, painting the town red. In the Soviet Union it is the color of revolution. In Christian nations it is the color of Satan, the hats of cardinals, the blood of the communion service. "Red," writes G. K. Chesterton ("The Red Town" in *Alarms and Discursions*), "is the most joyful and dreadful thing in the physical universe; it is the fiercest note, it is the highest light, it is the place where the walls of this world of ours wear thinnest and something beyond burns through."

No poet was ever more sensitive to the overtones of color words than Coleridge. Who doubts that he could have talked for hours about the "hooks and eyes of the memory" (as he once called the laws of association) by which the word "red" is linked with human experience? This word and its synonyms appear more frequently in *The Ancient Mariner* than any other color word. Miss Bodkin says she cannot read Coleridge's line about the ship's awful shadow without thinking of Dante's city of Dis, its red mosques glowing in a dark valley of the Inferno. She has given a valuable account of her own free associations and deeply felt emotions when she reads *The Ancient Mariner*—an account that will enrich any reader's understanding of the ballad—but Jung's collective unconscious is irrelevant to her central theme.

Miss Bodkin, as well as many later myth critics, often gives the impression that Coleridge's powerful elemental words and archetypal patterns entered his work unconsciously. This is hard to believe. Certainly the archetype of death and rebirth, in its Christian trappings, was deliberately and skillfully woven into the Mariner's tale. Young Coleridge, the son of a vicar, must have heard countless sermons on death and resurrection and the miraculous conversion of sinners. We know from his letters, written during the year that preceded the writing of his ballad, that he was intensely preoccupied with original sin, repentance, and the nature of the Fall and that his views were moving rapidly from Unitarianism back to orthodoxy. In "The Eolian Harp," a poem written two years before *The Ancient Mariner,* he speaks of himself as a "sinful and most miserable man" who had been healed and given peace by the saving mercies of the "Incomprehensible." Before writing his ballad he had planned an epic poem on the origin of evil. Most critics assume that this was his projected poem, "The Wanderings of Cain," which he did not finish because, as he tells us, he wrote *The Ancient Mariner* instead. It is impossible to suppose he could have written those stanzas about the mysterious wind that breathed on the Mariner without conscious awareness of the wind as a Biblical symbol of the Holy Spirit.

We can go further. *The Ancient Mariner* swarms with other religious symbols, not part of the rebirth archetype but so common in the religious literature and sermonizing of Coleridge's time that he could not have escaped recognizing them. How could he, for example, not have realized that the murder of the albatross carried emotional associations with the murder of Christ? We are told that the Polar Spirit "loved the bird that loved the man/Who shot him with his bow." No one with Coleridge's background and faith could fail to see here an analogy with God who loved His Son who loved the men who pierced him. The line just quoted is spoken by a daemon who a moment before had used the phrase, "By Him who died on cross." It is no accident, or upwelling of Coleridge's preconscious, that the albatross is hung on the Mariner's neck like a crucifix. Even the "cross" in "cross-bow" suggests the murder weapon with which Jesus was killed. I do not say that Coleridge worked out a vast, intricate, self-consistent metaphorical level for every stanza of his poem, or even that his Mariner was intended throughout as a symbol of all men, on the great sea voyage of life, burdened by Original Sin until they repent and are reborn. I do say that the Christian symbolism of the poem is so pervasive and obvious, the symbols so much a part of English Protestant culture in the late eighteenth century, that Coleridge would have had to have been simple-minded not to be aware of them in his ballad.

When we turn from the Christian theme of guilt and rebirth to subsidiary symbolic meanings of the ballad, we at once enter choppier waters. Coleridge did not leave a detailed analysis of his symbolic intent, and we can only speculate when we try to distinguish between the following three levels: (1) symbols consciously employed, (2) symbols unconsciously or semi-consciously used but nevertheless legitimate "meanings" of the poem, (3) symbols not intended in either of the above senses but which the poem has acquired almost by accident as the lines are given interpretations by later readers.

Of course there are no sharp lines separating the three levels. If you ask a poet whether he intended a certain symbolic meaning, he may answer, with complete honesty, that he does not know. He may have been dissatisfied with the awkward rhythms of a certain line. In altering the words he suddenly thought of an entirely new line of great verbal beauty. While writing it down he was startled to recognize that it has a symbolic meaning in harmony with the rest of the poem. Was this symbolic meaning "intended"? How can the poet himself say? It is consciously intended in the sense that, having written the line, he is aware of its symbolic meaning, approves of it, and lets the line stay. It is not consciously intended in the sense that he did not seek for such a symbol but only recognized it after he wrote it down. Was it unconsciously intended? Who can tell? It may have been a wholly accidental by-product of a search for a musical phrase; yet it certainly becomes a legitimate part of the poem's total meaning. Consider that well-known couplet of identical lines closing Robert Frost's "Stopping by Woods on a Snowy Evening":

> And miles to go before I sleep,
> And miles to go before I sleep.

It is difficult not to read the first line literally and its repetition as a symbolic reference to the Big Sleep. Yet we have Frost's own word for it that he did not intend this meaning when he wrote the poem, although he did admit once in a lecture that he had "the feeling" that his poem was "loaded with ulteriority." It is quite possible that Frost did not even unconsciously intend sleep as a symbol of death. But who would wish to eliminate this as one of the poem's meanings? Who would wish to discard the symbolic meanings that have been bestowed on passages in Homer? Every fantasy poem is so crowded with potential symbols that the probability of *some* accidental meanings, in harmony with the poem's central theme, is almost certain; but the difficulty of distinguishing this third level of meaning from the other two becomes very great indeed.

Physicists have a rule that if there are no laws to prevent something from happening in nature, it will. In poetic criticism there is a similar rule: If there is any possible attitude to take toward a great poem, some critic is sure to take it. *The Ancient Mariner* is no exception. It may be that no poem of comparable shortness has been subjected to so many varying symbolic interpretations.

Many critics view the poem as essentially a narrative about supernatural events that occur during a sea voyage, with few or no intended higher levels of meaning. True, Coleridge added an explicit moral at the end, almost as an afterthought, but (say these critics) he did not intend his ballad to carry the weight of vast symbolic meanings and little is gained in searching for them. It may be that here and there he intended a stanza to be symbolic, but on the whole, no higher metaphysical theme was part of the poet's intent. This is more or less the point of view of Lowes, Elisabeth Schneider (in *Coleridge, Opium and Kubla Khan,* 1953), Earl Leslie Griggs (in *The Best of Coleridge,* 1934), John Muirhead (in *Coleridge as Philosopher,* 1930), and many other critics.

At the other extreme are various attempts to treat the ballad as a carefully worked out allegory in which every character, object, and event is designed to carry a symbolic meaning that contributes to the higher theme. An early specimen of this approach is Gertrude Garrogues's paper, "Coleridge's 'Ancient Mariner,'" in the *Journal of Speculative Philosophy,* July, 1880. She regards the poem as throughout, stanza by stanza, a carefully planned allegory of the Christian theme of sin and redemption. Why does the Mariner stoppeth one of three? Because, as the Bible tells us, many are called but few are chosen; not every person is prepared to receive the story of salvation. Why is the sun above the mast at noon? Because the Mariner has finished the merry childhood of his voyage through life and has now reached maturity. And so on. The ballad is, thinks the author, the closest Coleridge ever came to writing that great work on Christian philosophy about which he talked so much!

There are many good things in Miss Garrogues's article, but, of course, she goes too far. I myself believe that the religious rebirth theme, in its Christian form, was consciously intended by Coleridge as the binding theme of his narrative—but surely not in a line-by-line way. It is likely that Coleridge himself was not fully aware of the extent of this theme when he first wrote the ballad. As Marius Bewley has said, only in "odd corners" of the poem can one feel fairly certain that the theme was intended. In his later changes, especially in the addition of the gloss, Coleridge sought to strengthen this theme and make it more apparent to the reader. But much of the action is hard to fit into any detailed allegory, and most critics who accept the poem as a Christian myth are careful not to press their metaphors too far. Like the Freudian symbol game, this religious symbol game is also easy to play.

One of the most perplexing problems, for readers who accept the sin-and-redemption theme, is to account for the apparent senselessness of the Mariner's crime. At the time Coleridge wrote the ballad he was well on his way toward the abandonment of Hartley's necessitarianism; there are many reasons to suppose that he intended the Mariner's lack of motive to dramatize an act that sprang directly from Original Sin. "A sin is an evil which has its ground or origin in the agent, and not in the compulsion of circumstances," he wrote many years later in *Aids to Reflection*. A man who commits a crime under the pressure of outer events is not really sinning in the deepest sense; he "may feel regret, but cannot feel remorse." Original Sin, he goes on to say, is a profound mystery which we cannot hope to understand. "It follows necessarily from the postulate of a responsible will. Refuse to grant this, and I have not a word to say. Concede this and you concede all."

If those were Coleridge's sentiments when he wrote his ballad, as I suspect they were, then the Mariner's motiveless cruelty may have a symbolic meaning essential to the poem's religious theme. The shooting of the albatross, like the shooting of President Kennedy, is banal and idiotic, more in keeping with Dante's imbecilic, three-headed Satan than with Milton's proud, handsome, understandable, and in some ways admirable archfiend. The really great sinner, Dante and Coleridge seem to be saying, is simply a fool. He wills his crime, knowing it a crime, but wills it for no particular reason.

C. M. Bowra, in his excellent chapter on *The Ancient Mariner* (in *The Romantic Imagination*, 1949), gives a well-balanced defense of the Christian theme. "We may begin by asking, as others have," he writes, "why there is all this 'pother about a bird,' but we end by seeing that, whatever the pother may be, it involves grave questions of right and wrong, of crime and punishment, and, no matter how much we enjoy the poetry, we cannot avoid being in some degree disturbed and troubled by it. Now this is surely the effect which Coleridge wished to produce. . . . The poem is a myth of a guilty soul and marks in clear stages the passage from crime through punishment to such redemption as is possible in this world."

Robert Penn Warren, a poet as well as a distinguished novelist and critic, also defends the religious theme, in his essay on *The Ancient Mariner*, "A Poem of Pure Imagination: An Experiment in Reading." (The essay first appeared in an edition of the ballad illustrated by Alexander Calder, but has since been reprinted, with revisions and additional notes, in Warren's *Selected Essays*.) It is the most influential analysis of *The Ancient Mariner* to have appeared in recent decades. Warren finds *two* metaphorical levels. The most obvious is the one we have been discussing; Warren calls it the theme of "sacramental vision" or "One life," thus emphasizing that the rebirth motif is linked with a vision of nature in which all living things are regarded as worthy of love. The secondary theme Warren calls the theme of the imagination. On this level the killing of the albatross is symbolic of a poet's crime against his imagination, for which he suffers a loss of creative power. I will not discuss this second theme because it would snare us in the complex topic of Coleridge's theory of imagination as distinct from fancy—and also because this aspect of Warren's essay has not met with general acceptance. His defense of Coleridge's primary theme, essentially the rebirth archetype of Maud Bodkin's analysis, is vigorous and carefully reasoned; his defense of the secondary theme is less convincing. For one thing, it involves an interpretation of the moon as a symbol of the imagination and the sun as a symbol of what Coleridge later called, under Kant's influence, the "understanding." Warren follows Kenneth Burke and George Herbert Clarke in regarding the moon as beneficent and the sun as malevolent; unfortunately, considerable mental gymnastics are required to explain such events as two hundred men falling dead under the moon (Part III) and the "sweet jargoning" of the angels under the sun (Part V).

Warren is considered a New Critic—a member of that loosely united group of writers who reacted against the Marxist political approach to literature during the thirties by directing attention back to the structure and intrinsic values of the work of art itself. Among detractors of this school, Warren's interpretation of *The Ancient Mariner* is a choice example of how easily New Critical enthusiasm for symbol hunting can lead one down dubious roads. The most vitriolic attack on Warren is Elder Olson's "Symbolic Reading of the Ancient Mariner." (It first appeared in a journal in 1948, was reprinted in *Critics and Criticism,* 1952, edited by Ronald S. Crane, and is now most accessible in *Visions and Revisions in Modern American Literary Criticism,* 1962 a paperback edited by Bernard S. Oldsey and Arthur O. Lewis, Jr.)

Olson, a poet and professor, belongs to the so-called Chicago school of criticism. This curious group, led by Crane and inspired by Aristotle and Richard P. McKeon, flourished in the forties at the University of Chicago, where its members developed a fairly elaborate program for criticism. It was not so much a new theory as an eclectic approach; it accepted all critical methods as legitimate but emphasized the need for a historical perspective and a special vocabulary, deriving from Aristotle, which the group believed to be more efficient than the vocabularies of rival schools. For some odd reason, however, when a member of the Chicago school actually practices criticism, he sounds just like any other critic, except for a more peevish tone and a stronger emphasis on the imbecility of anyone who disagrees with him. Olson has little use for either of Warren's themes, though he concentrates his fire on the secondary one, finding it compounded of "generous assumptions, undistributed middles, inconsistencies, misinterpretations, ignorationes elenchi, post hoc ergo propter hoc's, etc." That is supposed to take care of Warren—and also let one know that Olson has read Aristotle's discussion of logical fallacies.

Olson is so eager to defend the view that the primary end of a poem is to give "pleasure" that he appends an incredible footnote in which he says it is absurd to suppose an imitative poem *can* have a theme or meaning. "The words have a meaning," he writes; "they mean the poem; but why should the poem itself have any further meaning? What sense is there in asking about the meaning of something which is itself a meaning?" But hierarchies of meaning are commonplace. Marks on paper symbolize the word "stone," and the word "stone" symbolizes a small piece of rock. In the proverb, "A rolling stone gathers no moss," this bit of rock in turn symbolizes a person who drifts from place to place. Olson would no doubt reply that the sum of all such meanings is the proverb itself: therefore it is senseless to seek for a further meaning. Fair enough, but of course Warren can say exactly the same thing about a poem if "poem" is taken in this wide sense. One cannot say Olson is wrong but only that his terminology—at least in the footnote we are considering—departs so widely from common critical usage that needless confusion results.

There is, it seems to me, much less linguistic obfuscation if one accepts the utterly ordinary view that a narrative poem can have both literal and metaphorical levels of meaning. It is one thing to say that Coleridge did not intend a metaphorical level for his poem—that is a question to be decided by whatever evidence is available—but something else again to argue that the writer of a narrative poem cannot or should not consciously shape his story into a myth. For the right sort of reader, a metaphysical level of meaning can arouse as much "pleasure" as the imitative spectacle of a man experiencing fortunes and misfortunes. A narrative poem, like a person, can be a source of multiple pleasures; the metaphorical level no more weakens the pleasure aroused by the story itself, on its literal level, than a woman's intelligence makes her features less beautiful. Even when a poet's first intent is to arouse the kind of pleasure that derives from what the Chicago school persists in calling the "imitative" aspect of a poem, there is no reason why the poet cannot have all sorts of secondary motives, including the sale of the poem for money and the rhetorical motive of wishing to convert his readers to a certain point of view.

This brings us to a final question, one that has troubled critics for more than a century and a half: exactly what is the moral of *The Ancient Mariner?* Before trying to answer, we must first glance at the one document that bears most directly on the problem. In *Specimens of the Table Talk of the Late Samuel Taylor Coleridge,* as remembered by his nephew and son-in-law, Henry Nelson Coleridge, the following puzzling conversation occurs (May 31, 1830):

> Mrs. Barbauld once told me that she admired the *Ancient Mariner* very much, but that there were two faults in it—it was improbable, and had no moral. As for the probability, I owned that that might admit some question; but as to the want of a moral, I told her that in my own judgment the poem had too much; and that the only, or chief, fault, if I might say so, was the obtrusion of the moral sentiment so openly on the reader as a principle or cause of action in a work of such pure imagination. It ought to have had no more moral than the *Arabian Nights* tale of the merchant's sitting down to eat dates by the side of a well, and throwing the shells aside, and lo! a genie starts up, and says he *must* kill the aforesaid merchant *because* one of the date-shells had, it seems, put out the eye of the genie's son.[2]

The passage casts little light on the problem. First, we cannot be sure the conversation is recalled correctly. Assuming it is, we cannot be sure Coleridge was not pulling Mrs. Barbauld's leg. (Anna Letitia Barbauld was a popular poet, author, and writer of children's books; she was a devout Presbyterian, much given to pious, humorless moralizing.) Finally, assuming Coleridge did make these remarks and make them seriously, we cannot be sure just what he meant by them. By "moral," did he mean that quatrain beginning "He prayeth best," or was he referring to the theme of crime and punishment that is the framework of the poem? Perhaps he misunderstood Mrs. Barbauld. Did she have one meaning of "moral" in

---

2. A summary of the plot of the *Arabian Nights* tale (in which, by the way, the date shell or pit actually *kills* the genie's son) will be found in Humphry House's *Coleridge,* pp. 90–91, together with some observations on the tale's "moral" theme, without which, House argues, there would be no story.

mind and he another? It is amusing to read the conflicting ways in which critics have interpreted Coleridge's comment. The interpretation is always, of course, in support of the critic's way of viewing the poem. Those who find the religious theme either not there or of little significance cite Coleridge's remarks to Mrs. Barbauld in support of their view. The same passage is just as frequently cited by defenders of the religious theme, for did not the poet admit that there was "too much" of a moral in his ballad? The passage is too ambiguous to decide the matter. We will say no more about it.

There is no doubt, of course, that the poem closes with an explicitly stated moral in the "He prayeth best" stanza. On the literal level, it makes an obvious point. The Mariner's woes were brought about by his cruel killing of a bird. Had he loved the bird and not killed it, his shipmates would still be alive and he himself would not be doomed to wander about in a state of life-in-death, still doing penance for his crime. The moral tag is in keeping with the medieval atmosphere of the ballad, and Coleridge made no attempt to remove it from later printings. What are we to make of it?

What we make of it depends on whether we accept the symbolic religious theme. If we insist on reading the poem only on its literal level, we are likely to agree with Irving Babbitt (in an essay on Coleridge in *On Being Creative and Other Essays,* 1932) that the moral is something of a sham. There is, first of all, too "grotesque a disproportion between the mariner's initial act and its consequences." There is no serious ethical theme in the ballad, says Babbitt, "except perhaps a warning as to the fate of the innocent bystander; unless, indeed, one holds that it is fitting that, for having sympathized with the man who shot the albatross, 'four times fifty living men' should perish in torments unspeakable." And how is the Mariner relieved of this awful guilt? "By admiring the color of water snakes." Like Lamb, Babbitt dislikes all the miraculous elements of the poem. He sees them as the product of an abnormal mind, unduly preoccupied with the weird. It differs only in degree from one of Poe's horror tales and claims a "religious seriousness that at bottom it does not possess."

Lowes likewise finds that the moral, taken out of the poem's context, is untenable. "The punishment," he says, "measured by the standards of a world of balanced penalties, palpably does not fit the crime. But the sphere of balanced penalties is not the given world in which the poem moves. Within *that* world, where birds have tutelary daemons and ships are driven by spectral and angelic powers, consequence and antecedent are in keeping."

Yes, of course. If the poem is no more than a fantasy narrative, like an *Arabian Nights* tale, there is no reason why the Mariner should not warn his listener of the dangers of being cruel to birds. But from this viewpoint it is hard to escape Babbitt's feeling that the great nightmare voyage is finally climaxed by an utterly trivial piece of moralizing.

For the reader who is not repelled by the symbolic religious theme, the moral quatrain need no more be taken in such a literal, trivial sense than we need take the moral of *Moby Dick:* it is not good to make one's ultimate concern in life the killing of one particular whale. At the time Coleridge wrote his ballad he was deeply impressed by the sacramental view of nature as he found it in Hartley and in conversations with Wordsworth, who in turn had been influenced by Hartley. The concept that Albert Schweitzer calls "reverence for life" (note that "reverence to all things that God made and loveth" appears in the ballad's gloss) is expressed in other early poems by Coleridge. "The Raven," which Coleridge wrote in the same year that he wrote *The Ancient Mariner,* is in some ways, as Warren reminds us, a crude parallel of the ballad. And countless critics have pointed out that the "He loveth best" quatrain is surely a paraphrase of the following lines from Coleridge's "Religious Musings," written three years before the ballad:

> There is one Mind, one omnipresent Mind,
> Omnific. His most holy name is Love.
> Truth of subliming import! with the which
> Who feeds and saturates his constant soul,
> He from his small particular orbit flies
> With blest outstarting! From himself he flies,
> Stands in the sun, and with no partial gaze
> Views all creation; and he loves it all,
> And blesses it, and calls it very good!

If we are entitled to view the shooting of the albatross as a prototype of sin against God, we are entitled to interpret the word "small" in the moral quatrain as more than just a reference to birds and water snakes or the admonition that if one slaps a mosquito one should feel at least a twinge of remorse. The moral surely is— and it matters not a rap whether Coleridge did or did not consciously intend it this way—that one best loves God by loving his fellow man. "God be praised for all things!" Coleridge closed a letter in 1796, the year before he started his ballad. "A faith in goodness makes all nature good." Two weeks before he died, aware that he did not have long to live, Coleridge expressed regret that he would be unable to finish the systematic philosophy he had long hoped to write and added: "For, as God hears me, the originating, continuing and sustaining wish and design in my heart were to exalt the glory of His name; and, which is the same thing in other words, to promote the improvement of mankind. But *visum aliter Deo,* and 'His Will be done!'"

Need I remind some readers that it was Jesus who said that on two commandments hang all the Laws and Prophets? "Thou shalt love the Lord thy God with all thy heart, and with all thy soul, and with all thy mind, and with all thy strength: this is the first commandment. And the second is like, namely this, Thou shalt love thy neighbor as thyself. There is none other commandment greater than these." (Mark 12:30,31.) In Coleridge's ballad this moral may have a jingly, Sunday school sound, as well as grotesque associations with albatrosses and water snakes, but there is no reason why we should not take it, on the higher mythic level of the poem, in the widest sense.

Many critics have defended the simplicity and naïveté of the ballad's moral quatrain, but none has done so more effectively than Mrs. Margaret Oliphant. In her *Literary History of England* (1882, Vol. I, Chap. 7, "The Lyrical Ballads"), she writes: "And then comes the ineffable, half-childish, half divine simplicity of those soft moralizings at the end, so strangely different from the tenor of the tale, so wonderfully perfecting its visionary strain. After all, the poet seems to say, after this weird excursion into the very deepest, awful heart of the seas and mysteries, here is your child's moral, a tender little half-trivial sentiment, yet profound as the blue depths of heaven:

> He prayeth best, who loveth best
> All things both great and small;
> For the dear God who loveth us,
> He made and loveth all.

"This unexpected gentle conclusion brings our feet back to the common soil with a bewildered sweetness of relief and soft quiet after the prodigious strain of mental excitement which is like nothing else we can remember in poetry. The effect is one rarely produced, and which few poets have the strength and daring to accomplish, sinking from the highest notes of spiritual music to the absolute simplicity of exhausted nature."

At present, Christian churches in this country are suddenly discovering the moral's application to the racially "small" in our midst, our black minority. "He prayeth best who loveth best." How fare, one wonders, the prayers of our southern Catholics and Protestants who refuse to take communion if the person next to them has skin of a different color? Cross bows come in all shapes and sizes.

I should like to end my remarks with an ambiguous fable of my own—or, rather, a fable I discovered in *The New York Times*—Sunday, February 2, 1964. According to the Times and later news releases from the National Audubon Society, the U.S. Navy had found it necessary to destroy about 20,000 of an estimated 150,000 albatrosses that nest on Sand and Eastern islands, part of the Midway group in the Hawaiian archipelago.

Two species of albatross—the Laysan and the blackfooted albatrosses—build nests on the island. The huge birds have a habit of getting in the way of Navy planes when the planes take off. No Navy personnel had yet been killed, but many planes had been damaged by colliding with the birds. One plane had its radar equipment and three rudders knocked off. Every hour and a half a giant plane takes off, with twenty-two men and six million dollars worth of electronic equipment. Every hour and a half a plane lands. The flights are part of the country's early-warning radar network.

The Navy had tried various measures. It set off flares, mortar shells, and bazookas. It shouted at the birds through loudspeakers. It hoisted scarecrows. The friendly gooney birds (as they are called by the local sailors) seemed to enjoy watching these antics. The Navy destroyed the nests. The goonies made new ones. Finally, more drastic steps had to be taken. The birds show no fear of man, so it was easy to capture them, put them in sealed chambers, and kill them with carbon monoxide from Navy truck exhausts. Quite apart from all this, about one hundred albatrosses are killed each week by flying into a huge antenna of guy wires on Eastern Island.

Carl W. Buchheister, president of the National Audubon Society, had just returned from Midway where he had gone to investigate the situation. He recommended hiring a full-time resident ornithologist, continuing research on ways of keeping the birds off the airstrips, and flagging the guy wires. The Navy, he reported, shared his "extreme regret" that their "elimination program had become necessary."

The Mighty Casey

One of the most humiliating defeats in the history of the New York Yankees took place on Sunday, October 6, 1963. Because a well-thrown ball bounced off the wrist of first baseman Joe Pepitone, the Yanks lost the fourth straight game and the World Series to their old enemies, the former Brooklyn (but by then Los Angeles) Dodgers. Across the top of next morning's *New York Herald Tribune* ran the headline: "The Mighty Yankees Have Struck Out." Lower on the same page another headline read: "But There's Still Joy in Mudville." (The New York Stock Exchange was holding up well under the grim news.)

Every reader of those headlines knew that they came straight out of that immortal baseball ballad, that masterpiece of humorous verse, *Casey at the Bat.* Not one in ten thousand could have named the man who wrote that poem.

His name was Ernest Lawrence Thayer. The story of how young Thayer, at the age of twenty-five and fresh out of Harvard, wrote *Casey*—and of how the ballad became famous—has been told before. But it has seldom been told accurately or in much detail, and, in any case, it is worth telling again.

Thayer was born in Lawrence, Massachusetts, on August 14, 1863, exactly one hundred years before the mighty Yankees made their celebrated strike out. By the time he entered Harvard, the family had moved to Worcester, where Edward Davis Thayer, Ernest's well-to-do father, ran one of his several woolen mills. At Harvard, young Thayer made a brilliant record as a major in philosophy. William James was both his teacher and friend. Thayer wrote the annual Hasty Pudding play. He was a member of the Delta Kappa Epsilon fraternity and the highly exclusive Fly Club. He edited the Harvard *Lampoon,* the college's humor magazine. Samuel E. Winslow, captain of the senior baseball team (later he became a congressman from Massachusetts), was young Thayer's best friend. During his last year at Harvard, Thayer never missed a ball game.

Another friend of Thayer's college years was the *Lampoon's* business manager, William Randolph Hearst. In 1885, when Thayer was graduated *magna cum laude*—he was Phi Beta Kappa and the Ivy orator of his class—Hearst was unceremoniously booted off the Harvard Yard. (He had a habit of playing practical jokes that no one on the faculty thought funny, such as sending chamber pots to professors, their names inscribed thereon.) Hearst's father had recently bought the ailing *San Francisco Examiner* to promote his candidacy as United States senator from California. Now that young Will was in want of something to occupy his time, the elder Hearst turned the paper over to him.

Thayer, in the meantime, after wandering around Europe with no particular goal, settled in Paris to brush up on his French. Would he consider, Hearst cabled him, returning to the United States to write a humor column for the *Examiner's* Sunday supplement? To the great annoyance of his father, who expected him to take over the American Woolen Mills someday, Thayer accepted Hearst's offer.

Thayer's contributions to the paper began in 1886. Most were unsigned, but starting in October 1887 and continuing into December he wrote a series of ballads that ran in the Sunday editions, about every other week, under the by-line of "Phin." (At Harvard his friends had called him Phinny.) Then ill health forced him to return to Worcester. He continued for a while to send material to the *Examiner,* including one final ballad, *Casey.*[1] It appeared on Sunday, June 3, 1888, page 4, column 4, sandwiched inconspicuously between editorials on the left and a weekly column by Ambrose Bierce on the right.

No one paid much attention to *Casey.* Baseball fans in San Francisco chuckled over it and a few eastern papers reprinted it, but it could have been quickly forgotten had it not been for a sequence of improbable events. In New York City a rising young comedian and bass singer, William De Wolf Hopper, was appearing in

---

1. In an interview with Homer Croy, Thayer is quoted as saying that in the fall of 1887 he had been reading W. S. Gilbert's *Bab Ballads* and that this had prompted him to attempt similar ballads for his newspaper column. *Casey* was written, Thayer said, in May 1888. He received five dollars for each ballad.

Prince Methusalem, a comic opera at Wallack's Theatre, at Broadway and 30th Street. One evening (the exact date is unknown; it was probably late in 1888 or early in 1889)[2] James Mutrie's New York Giants and Pop Anson's Chicago White Stockings were invited to the show as guests of the management. What could he do on stage, Hopper asked himself, for the special benefit of these men? I have just the thing, said Archibald Clavering Gunter, a novelist and friend. He took from his pocket a ragged newspaper clipping that he had cut from the *Examiner* on a recent trip to San Francisco. It was *Casey.*

This, insisted Gunter, is great. Why not memorize it and deliver it on stage? Hopper did exactly that, in the middle of the second act, with the Giants in boxes on one side of the theatre, the White Stockings in boxes on the other. This is how Hopper recalled the scene in his memoirs, *Once a Clown Always a Clown:*

> When I dropped my voice to B flat, below low C, at "the multitude was awed," I remember seeing Buck Ewing's[3] gallant mustachios give a single nervous twitch. And as the house, after a moment of startled silence, grasped the anticlimactic dénouement, it shouted its glee.
>
> They had expected, as any one does on hearing Casey for the first time, that the mighty batsman would slam the ball out of the lot, and a lesser bard would have had him do so, and thereby written merely a good sporting-page filler. The crowds do not flock into the American League parks around the circuit when the Yankees play, solely in anticipation of seeing Babe Ruth whale the ball over the centerfield fence. That is a spectacle to be enjoyed even at the ex-

2. In his memoirs, Hopper gives the date as May 13, 1888. This is certainly incorrect because *Casey* was not printed in the *San Francisco Examiner* until June 3 of that year. Hopper also wrongly recalls that the initials "E.L.T." were appended to the ballad. In *Famous Single Poems,* Burton Stevenson says he received a letter from Hopper correcting the date given in his memoirs and stating his conviction that the historic first recitation of *Casey* was in August, 1888. Thanks to the diligent research of Jules L. Levitt, of Binghamton, New York, this has now been verified. A review in *The New York Times,* August 15, 1888, page 4, describes the memorable occasion on the night of August 14 when Hopper gave his first recitation of *Casey,* and how it was "uproariously received" by the audience.

3. William ("Buck") Ewing, catcher for the New York Giants. He is said to have been the first catcher to throw to second without wasting time by standing up. On one famous occasion he stole second, then third, and shouted out that he intended to steal home, which he did. Robert Smith, in his picture book *Baseball's Hall of Fame* (Bantam, 1965), says that a lithograph depicting Ewing's mighty slide, as he stole home, was widely sold all over New York City. In 1883 Buck led the National League in home runs.

pense of the home team, but there always is a chance that the Babe will strike out, a sight even more healing to sore eyes, for the Sultan of Swat can miss the third strike just as furiously as he can meet it, and the contrast between the terrible threat of his swing and the futility of the result is a banquet for the malicious, which includes us all. There is no more completely satisfactory drama in literature than the fall of Humpty Dumpty.

Astonished and delighted with the way his audience responded to *Casey*, Hopper made the recitation a permanent part of his repertoire. It became his most famous bit. Wherever he went, whatever the show in which he was appearing, there were always curtain shouts for "Casey!" By his own count he recited it more than 10,000 times, experimenting with hundreds of slight variations in emphasis and gesture to keep his mind from wandering. It took him exactly five minutes and forty seconds to deliver the poem.[4]

"When my name is called upon the resurrection morning," he wrote in his memoirs, "I shall, very probably, unless some friend is there to pull the sleeve of my ascension robes, arise, clear my throat and begin: 'The outlook wasn't brilliant for the Mudville nine that day.'" The poem, declared Hopper, is the only truly great comic poem written by an American. "It is as perfect an epitome of our national game today as it was when every player drank his coffee from a mustache cup. There are one or more Caseys in every league, bush or big, and there is no day in the playing season that this same supreme tragedy, as stark as Aristophanes for the moment, does not befall on some field. It is unique in all verse in that it is not only funny and ironic, but excitingly dramatic, with the suspense built up to a perfect climax. There is no lame line among the fifty-two."

Let us pause for some moments of irony. Although Hopper was famous in his day as a comic opera star, today he is remembered

---

4. Hopper's deep, rich voice, reciting *Casey*, was first recorded in 1906 on a Victor Grand Prize Record, No. 31559. This was reissued in 1913 as No. 35290, with a reverse side bearing *The Man Who Fanned Casey*, recited by Digby Bell, a popular singing comedian of the day. A subsequent recording of Hopper doing *Casey* was released by Victor in 1926 as "orthophonic recording" No. 35783. On the flip side Hopper recited the parody, *O'Toole's Touchdown*. Perhaps Hopper made other recordings of *Casey*, but these three are all I could find. A garbled version of *Casey*, cluttered with sound effects and corny music, was recorded much later by Lionel Barrymore on two sides of an M.G.M. record. A recording by Mel Allen is on a Golden Record for children.

for three things: (1) Hedda Hopper was the fifth of his six wives, (2) William Hopper, his only child by Hedda, played Paul Drake of the "Perry Mason" TV show, and (3) he was the man who recited *Casey.*

More ironic still, Gunter—who wrote thirty-nine novels including a best seller called *Mr. Barnes of New York*—has found his way into terrestrial immortality only because he happened to take *Casey* out of a newspaper and pass it on to Hopper. We must not belittle this achievement. "It is easy enough to recognize a masterpiece after it has been carefully cleaned and beautifully framed and hung in a conspicuous place and certified by experts," wrote Burton Stevenson, a critic and poetry anthologist, with specific reference to Gunter and *Casey.* "But to stumble over it in a musty garret, covered with dust, to dig it out of a pile of junk and know it for a thing of beauty—only the connoisseur, can do that."

Gunter was the connoisseur, but Hopper made the poem famous. All over the United States, newspapers and magazines began to reprint it. No one knew who "Phin" was. Editors either dropped the name altogether or substituted their own or a fictitious one. Stanzas were lost. Lines got botched by printers or rewritten by editors who fancied themselves able to improve the original. Scarcely two printings of the poem were the same. In one early reprinting, by the *New York Sporting Times,* July 29, 1888, Mudville was changed to Boston and Casey's name to Kelly, in honor of Mike ("King") Kelly, a famous Chicago star who had recently been bought by the Boston team.

After the banquet, at a Harvard decennial class reunion in 1895, Thayer recited *Casey* and delivered an eloquent speech, tinged with ironic humor and sadness. (It is printed, along with *Casey,* in *Harvard University, Class of 1885: Secretary's Report No. V,* 1900, Pp. 88–96.) The burden of his address was that the world turns out to be not quite the bowl of cherries that a haughty Harvard undergraduate expects it to be. Surely the following passage is but a roundabout way of saying that it is easy to strike out:

> We give today a wider and larger application to that happy phrase of the jury box, "extenuating circumstances." We have found that playing the game is very different from watching it played, and that splendid theories, even when accepted by the combatants, are apt to be lost sight of in the confusion of active battle. We have reached an age, those of us to whom fortune has assigned a post in life's struggle, when, beaten and smashed and biffed by the lashings of the dragon's tail, we begin to appreciate that the old man was not such a damned fool after all. We saw our parents wrestling with that same dragon, and we thought, though we never spoke the thought aloud, "Why

don't he hit him on the head?" Alas, comrades, we know now. We have hit the dragon on the head and we have seen the dragon smile.

From time to time various "Caseys" who actually played baseball in the late 1880s claimed to have been the inspiration for the ballad. But Thayer emphatically denied that he had had any ball player in mind for any of the men mentioned in *Casey*. When the *Syracuse Post-Standard* wrote to ask him about this, he replied with a letter that is reprinted in full in Lee Allen's entertaining book on baseball, *The Hot Stove League:*

> The verses owe their existence to my enthusiasm for college baseball, not as a player, but as a fan. . . . The poem has no basis in fact. The only Casey actually involved—I am sure about him—was not a ball player. He was a big, dour Irish lad of my high school days. While in high school, I composed and printed myself a very tiny sheet, less than two inches by three. In one issue, I ventured to gag, as we say, this Casey boy. He didn't like it and he told me so, and, as he discoursed, his big, clenched, red hands were white at the knuckles. This Casey's name never again appeared in the *Monohippic Gazette.* But I suspect the incident, many years after, suggested the title for the poem. It was a taunt thrown to the winds. God grant he never catches me.

By 1900 almost everyone in America had heard or read the poem. No one knew who had written it. For years it was attributed to William Valentine, city editor of the *Sioux City Tribune,* Iowa. One George Whitefield D'Vys, of Cambridge, actually went about proudly proclaiming himself the author; he even signed a document to this effect and had it notarized. In 1902 *A Treasury of Humorous Poetry,* edited by Frederic Lawrence Knowles, credited the poem to someone named Joseph Quinlan Murphy. To this day no one knows who Murphy might have been, if he really existed, or why Knowles supposed he had written *Casey.*

Hopper himself did not find out who wrote the ballad until about five years after he began reciting it. One evening, having delivered the poem in a Worcester theatre, he received a note inviting him to a local club to meet *Casey's* author. "Over the details of wassail that followed," Hopper wrote later, "I will draw a veil of charity." He did disclose, however, that the club members had persuaded Thayer himself to stand up and recite *Casey*. It was, Hopper declared, the worst delivery of the poem he had ever heard. "In a sweet, dulcet Harvard whisper he [Thayer] implored Casey to murder the umpire, and gave this cry of mass animal rage all the emphasis of a caterpillar wearing rubbers crawling on a velvet carpet."

Thayer remained in Worcester for many years, doing his best to please his father by managing one of the family mills. He kept quietly to himself, studying philosophy in spare hours and reading classical literature. He was a slightly built, soft-spoken man, inclined to deafness in his middle years (he wore a hearing aid), always gracious, charming, and modest. Although he dashed off four or five more comic ballads in 1896, for Hearst's *New York Journal,* he continued to have a low opinion of his verse.

"During my brief connection with the *Examiner,*" Thayer once wrote, "I put out large quantities of nonsense, both prose and verse, sounding the whole newspaper gamut from advertisements to editorials. In general quality *Casey* (at least in my judgment), is neither better nor worse than much of the other stuff. Its persistent vogue is simply unaccountable, and it would be hard to say, all things considered, if it has given me more pleasure than annoyance. The constant wrangling about the authorship, from which I have tried to keep aloof, has certainly filled me with disgust." Throughout his life Thayer refused to discuss payments for reprintings of *Casey.* "All I ask is never to be reminded of it again," he told one publisher. "Make it anything you wish."

Never happy with the woolly details of the family mills, Thayer finally quit working for them altogether. After a few years of travel abroad, he retired in 1912 to Santa Barbara, California. The following year—he was then fifty—he married Mrs. Rosalind Buel Hammett, a widow from St. Louis. They had no children.

Thayer remained in Santa Barbara until his death in 1940. Friends said that toward the end of his life he softened a bit in his scornful attitude toward *Casey.* By then even English professors, notably William Lyon Phelps of Yale, had hailed the poem as an authentic native masterpiece. "The psychology of the hero and the psychology of the crowd leave nothing to be desired," wrote Phelps in *What I Like in Poetry* (Scribner's, 1934). "There is more knowledge of human nature displayed in this poem than in many of the works of the psychiatrist. Furthermore, it is a tragedy of Destiny. There is nothing so stupid as Destiny. It is a centrifugal tragedy, by which our minds are turned from the fate of Casey to the universal. For this is the curse that hangs over humanity—our ability to accomplish any feat is in inverse ratio to the intensity of our desire."

Thayer attended a class reunion at Harvard in 1935. Friends reported that he was visibly touched when he saw a classmate carrying a large banner that read: "An '85 Man Wrote *Casey!*"

Music for Thayer's poem was written by Sidney Homer and published by G. Schirmer, New York City, in 1920. (The sheet music bears the general title: *Six Cheerful Songs to Poems of American Humor.* Casey is No. 3.) Two silent movies were about Casey. The first starred Hopper himself as the mighty batsman. It was produced by Fine Arts–Triangle and released June 22, 1916. (Scenes from this film may be found in *The Triangle,* Vol. 2, June 17, 1916.) A remake, with Wallace Beery in the leading role (supported by Ford Sterling and Zasu Pitts), was released by Paramount on April 17, 1927. I can still recall Beery, bat in one hand and beer mug in the other, whacking the ball so hard that an outfielder had to mount a horse to retrieve it. An animated cartoon of the famous strikeout was included in Walt Disney's 1946 release, *Make Mine Music,* with Jerry Colonna providing an off-camera recitation of Thayer's ballad. (Since 1960 this has been available as a reissued short feature from Encyclopedia Britannica Films). In 1953 Disney released a cartoon short called *Casey Bats Again.* It tells how Casey organized a girls' baseball team, then, to save the game in a pinch, dressed like a girl and batted in the winning run.

The most important continuation and elaboration of the Casey story is an opera, *The Mighty Casey,* which had its world premiere in Hartford, Connecticut, on May 4, 1953.[5] William Schuman, who wrote the music, is now the president of New York City's Lincoln Center for the Performing Arts. He has been a baseball buff since his childhood on New York's upper west side. In his teens he seriously considered becoming a professional ball player. "Baseball was my youth," he has written. "Had I been a better catcher, I might never have become a musician." But in his early twenties his love of music won out, and by 1941 (he was then thirty-one) his *Third Symphony* lifted him into the ranks of major United States composers. From 1935 to 1961 he was president of the Juilliard School of Music, and since 1962 he has been head of Lincoln Center. Jeremy Gury, who wrote *The Mighty Casey's* libretto, has been senior vice-president and creative director of Ted Bates & Company, New York City, since 1953. Before he entered advertising he had been managing editor of *Stage Magazine.* He has written a number of childrens' books (*The Round and Round Horse, The Wonderful World*

5. This production, with Louis Venora as the mute Casey, was by the Julius Hartt Opera Guild. In addition to *The New York Times* review mentioned later, see also reviews in Time (vol. 60, May 18, 1953, p. 61) and *Musical America* (vol. 73, June, 1953).

of Aunt Trudy, and others) and one play (with music by Alex North), *The Hither and Thither of Danny Dither.*

The Mighty Casey obviously is the product of two knowledge-able baseball enthusiasts. They have expanded the Casey myth with such loving insight, such full appreciation of the nuances in Thay-er's ballad, that no Casey fan need hesitate to add the opera to the *Casey* canon. It is sad that Thayer did not live to see it. The details of its plot mesh so smoothly with the poem that one feels at once, "Yes, of course, that *must* have been the way it happened."

Mudville is playing Centerville for the state championship of the Inter-Urban League. In the bleachers, watching the crucial game, are two big-league scouts. Casey's girlfriend, Merry, knows that if Casey does well in the game he will leave Mudville forever; yet she loves him enough to offer up a prayer, in the last half of the ninth, that Flynn and Blake will not prevent her hero from coming to bat. While the fateful half is enacted in slow pantomime, the Watchman of the ballpark recites Thayer's entire poem—alas, a cor-rupted version, but it does include two new quatrains by Gury. The final pitch is made in slow motion, an ominous drumroll beginning as soon as Fireball Snedeker (how could the Centerville pitcher have been named anything else?) releases the leather-covered sphere. Casey's tragic swing creates a monstrous wind that blows back the crowd in the grandstand, while a great whining sound from the orchestra fades off into deathlike silence. The crowd, like a Greek chorus, sings "Oh, Somewhere"—the poem's final stanza—as Casey slowly exits. Throughout the entire opera—it runs about an hour and twenty minutes—Casey speaks not a word. "We simply felt," the authors explain in their libretto, "that one so god-like should not speak. The magnificence of Casey is above mere words."

The Mighty Casey has yet to have a full-scale production in New York City. (It is not easy to put on a short opera that calls for a forty-piece orchestra and a chorus of fifty voices!) After its one per-formance in Hartford, there was a CBS television production of *The Mighty Casey* on the "Omnibus" show, March 6, 1955,[6] and it has been performed by small companies in San Francisco, Annapolis, and elsewhere. There have been several productions in baseball-loving Japan. Harold C. Schonberg, reviewing the Hartford produc-

6. The "Omnibus" show featured Danny Scholl as Casey, Elise Rhodes as Merry. A preview, with pictures, appeared in the *New York Herald Tribune,* March 4, 1955. Harold C. Schonberg reviewed the "Omnibus" show for *The New York Times,* March 7, 1955.

tion in *The New York Times,* (May 5, 1953, P. 34), spoke of the music as "lively, amusing, tongue-in-cheek." He felt that Schuman's "dry, often jerky melodic line with all its major sevenths and ninths, his austere harmonies and his rhythmic intensity," does not quite fit Thayer's "pleasant little fable." Can it be that the music critic of *The New York Times* is not a baseball fan? Pleasant little fable, indeed! *Casey* is neither pleasant nor little, it is tragic and titanic. Perhaps Schuman's intense music is not so inappropriate after all.

Several flimsy paperback copies of the poem, with illustrations, were printed around the turn of the century, but it was not until 1964 that *Casey* appeared in handsomely illustrated hardcover editions. I have yet to see two printings of the poem exactly alike. The Franklin Watts 1964 book comes closer to the original than any currently available printing; it follows the first version word for word except for the correction of two obvious printer's errors and cleaner punctuation here and there.

How can one explain *Casey's* undying popularity? It is not great poetry. It was written carelessly. Parts of it are certainly doggerel. Yet it is almost impossible to read it several times without memorizing whole chunks, and there are lines so perfectly expressed, given the poem's intent, that one cannot imagine a word changed for the better. T. S. Eliot admired the ballad and even wrote a parody about a cat, *Growltiger's Last Stand,* in which many of Thayer's lines are echoed.[7]

---

7. This ballad (in Eliot's *Old Possum's Book of Practical Cats,* Harcourt, Brace and Co., 1939) relates the fall of the great, one-eyed pirate cat, Growltiger, "The Terror of the Thames." Growltiger pursues his evil aims by roaming up and down the river on a barge. But one balmy moonlit night,when his barge is anchored at Molesey and his raffish crew members are either asleep or wetting their beards at nearby pubs, he is cornered by a gang of Siamese cats and, to his vast surprise, forced to walk the plank:

> He who a hundred victims had driven to that drop,
> At the end of all his crimes was forced to go ker-flip, ker-flop.

The ballad's fourteen stanzas follow the rhyme scheme and iambic septameter of *Casey.* The final stanza begins: "Oh there was joy in Wapping . . ." (Wapping, on the Thames, is a dreary dock section of Stepney, an eastern borough of London. Its inhabitants—mostly longshoremen, sailors, and factory hands—are called Wappingers. Boswell writes of an occasion on which Samuel Johnson talked about "the wonderful extent and variety of London, and observed, that men of curious inquiry might see in it such modes of life as very few could even imagine. He in particular recommended to us to *explore Wapping* . . ." This Boswell did. But he adds: ". . . whether from that uniformity which has in modern times, in a great degree, spread through

The poem's secret can be found, of all places, in the auto-biography of George Santayana, another famous Harvard philosopher. Santayana was one of Thayer's associate editors on the *Lampoon*. "The man who gave the tone to the *Lampoon* at that time," Santayana writes, "was Ernest Thayer. . . . He seemed a man apart, and his wit was not so much jocular as Mercutio-like, curious and whimsical, as if he saw the broken edges of things that appear whole. There was some obscurity in his play with words, and a feeling (which I shared) that the absurd side of things is pathetic. Probably nothing in his later performance may bear out what I have just said of him, because American life was then becoming unfavorable to idiosyncrasies of any sort, and the current smoothed and rounded out all the odd pebbles."[8]

But Santayana was wrong. One thing *did* bear this out, and that was *Casey*. It is precisely the blend of the absurd and the tragic that lies at the heart of Thayer's remarkable poem. Casey is the giant of baseball who, at his moment of potential triumph, strikes out. A pathetic figure—yet comic because of the supreme arrogance and confidence with which he approached the plate.

> There was ease in Casey's manner as he stepped into his place;
> There was pride in Casey's bearing and a smile on Casey's face.
> And when, responding to the cheers, he lightly doffed his hat,
> No stranger in the crowd could doubt 'twas Casey at the bat.

It is the shock of contrast between this beautiful build up and the final fizzle that produces the poem's explosion point. The story of Casey has become an American myth because Casey is the incomparable, towering symbol of the great and glorious poop-out.

One might argue that Thayer, with his extraordinary beginning at Harvard, his friendship with James and Santayana, his life-long immersion in philosophy and the great books, was himself something of a Casey. In later years his friends were constantly urging him to write, but he would always shake his head and reply, "I have nothing to say." Not until just before his death, at the age of seventy-seven, did he make an attempt to put some serious

---

every part of the Metropolis, or from our want of sufficient exertion, we were disappointed.")

8. Santayana, *Persons and Places* (Scribner's, 1943), page 197. Santayana's failure to mention *Casey* may be accounted for, in part, by the fact that he greatly preferred football to baseball.

thoughts on paper. Then it was too late. "*Now* I have something to say," he said, "and I am too weak to say it."[9]

But posterity's judgments are hard to anticipate. Thayer's writing career was no strike out. He swatted one magnificent home run, *Casey at the Bat;* and as long as baseball is played on this old earth, on Mudville, the air will be shattered over and over again by the force of Casey's blow.

9. These remarks of Thayer's are quoted in his obituary in the *Santa Barbara* (California) *News-Press,* August 22, 1940.

The Martian Chronicles

Science fiction, unlike other fiction, is vulnerable to a peculiar kind of malady. A science-fiction novel or short story, even a basic plot theme, can be rendered inoperative by a single scientific discovery. Occasionally it may work the other way. Some old, forgotten tale suddenly becomes "prophetic" in the light of a new discovery and enjoys a temporary revival. Even when this happens, the same discovery is likely to send fifty other stories down the drain.

Consider the sad fate of Edgar Rice Burroughs's Martian novels. What great adventures they were in the twenties and thirties! Without Burroughs, Ray Bradbury has declared, "*The Martian Chronicles* would never have been born. Lacking refinement, with exquisite vulgarity, he pummeled and shoved me into the field of writing, where I collided with the better minds of Huxley and Wells along the way. But Burroughs was first and foremost the vulgarian who took me out under the stars of Illinois and pointed up and said, with John Carter, simply: Go There. So, finally in my twenties, I went."

Unfortunately, much of the excitement generated by Burroughs's Mars books rested on the possibility, remote but genuine, that the surface of ruddy old Barsoom, crisscrossed by dying canals, actually teemed with humanoid races and unearthly creatures. The barren photographs of the Mariner space probes killed all that. Are the Martian novels of Burroughs rich enough in other values to survive this blow? One suspects not. His Tarzan novels can still be enjoyed precisely because most Tarzan buffs know less about Africa than about Mars. The Tarzan books swarm with as many howlers as

I had the privilege of writing this article as an introduction to an edition of *The Martian Chronicles*, published in 1974 by The Heritage Press for their Limited Editions Club. It is reprinted here with permission.

Burroughs's Mars books. But if the reader is not aware of them, what do they matter?

In the light of the cold, computer-strengthened photographs of the Mariner probes, Bradbury's Tyrr, let it be said at once, is as quaint and obsolete as old Barsoom. There are no canals on Mars. There is not even one open body of water on Mars. The planet's attenuated atmosphere is mostly carbon dioxide, its oxygen content too thin to permit breathing by anyone from Earth. The greening of Tyrr that takes place in 2001, when the trees and grass planted by Benjamin Driscoll burst from the black Martian soil, could not have helped. The low Martian gravity (about two-fifths that of Earth) would not prevent plant-generated oxygen from evaporating into space. Nor does Bradbury consider the effect of Mars's weak gravity on the behavior of the settlers.

In the second chronicle, Ylla sees the two white moons of Mars rise together over the desert. Alas, Phobos whirls around the planet faster than the planet rotates. On Mars, Deimos rises in the east, Phobos in the west. Speaking at a symposium at Caltech a few years ago, Bradbury recalled how a nine-year-old boy had once informed him of this peculiarity of Phobos. "So I hit him," said Bradbury. "I'll be damned if I'll be bullied by bright children!"

There is no way, of course, to revise the *Chronicles* so they conform to what even now is known about Mars. And this raises a fascinating question. How has it happened that Bradbury's stories about Mars, written originally for science-fiction magazines and later spliced together to make a sort of novel, have survived the malady of scientific progress? Why is it that reading the *Chronicles* today is as rewarding an experience as it ever was, perhaps more so?

To answer this we must begin with an often-stated fact: Bradbury never intended his Mars stories to be science fiction in the usual sense. He did not try to write realistic science stories in the manner of Jules Verne or H. G. Wells. He did not try to write romantic science fiction in the manner of Edgar Rice Burroughs. What he wrote is fantasy lightly touched by science. His Tyrrian mythology is as remote from Mars as the mythology of Mount Olympus is remote from the actual isles of ancient Greece.

There is not *one* Mars in Bradbury's mythology, there are three. First, there is the Mars that flourished before the coming of the earthmen. It is Bradbury's Oz. It is a dream utopia, a race of wise and beautiful people—telepathic, clairvoyant, precognitive, golden-eyed—who live in glass and crystal cities that gleam like carved chessmen, cities as fragile as the glass city in *Dorothy and the Wizard in Oz*. Earth has become the anti-utopia Bradbury details elsewhere: in his novella and play, *Fahrenheit 451,* and in numerous short stories. A too-rapidly advancing science and technology has crushed the people of earth into "the tubes, tins, and boxes" of ugly, noisy, polluted cities, taken away their freedoms, stuffed them with hate, and handed them atom bombs to play with.

Next, there is the colonized Mars. The microbes of a ridiculous children's disease, carried by earth's astronauts, have killed all but a small remnant of the natives. The planet is under the full control of settlers. The graceful Martian culture has been obliterated as decisively as Rome obliterated Carthage, as thoroughly as America obliterated the cultures of the Indians. Stories about the colonization of Mars are stories about lonely pioneers, the smell of rockets (as Bradbury put it in a Mars story not in this book) replacing the smell of buffalo. Behind the fantasy and the new geographical names are the same old heroisms, prejudices, savageries, and cultural shocks, the same rebellions and assimilations that always accompany the movements of people into alien lands.

And last, there is the Mars that may arise after the October of the final chronicle. The planet's first settlers have gone back to Mother Earth, to participate in the raging wars. Earth is almost destroyed. A man escapes from ravaged Earth and takes his family to desolated Mars (soon others will follow). One of his sons picks a dead Martian city to be their home. Will these new Martians do better than their predecessors? As always, mankind has escaped by the skin of its teeth. The future is unknown and perilous, but not without hope.

"Old Mars," Bradbury cries out in one of his poems, "then be a hearth to us." What does it matter if the ancient dreams of native Martian cities "self-destruct"? Mars will be our stopping place, a temporary nest before we move on to unimaginable destinies on planets that circle other suns. *We* will be the native Martians. Our children and our children's children may yet put down on some actual map the glittering cities of a new Barsoom.

So much for the *Chronicles'* leading themes. But there is more, much more, to the book than this. Almost all of Bradbury's other major themes, dramatized in many of his stories and novels, are here. We come upon them quietly, in the strange lights and shadows of Tyrr—themes as untouchable by space probes and future landings as the themes of *A Midsummer Night's Dream*.

There are the mystery of time and the sadness of the irretrievable past. These emotions are at the heart of many other Bradbury stories—"A Scent of Sarsaparilla," "The Lake," "The Tombling Day," most of the episodes in *Dandelion Wine*. Who but Bradbury could have used such emotions, as he does in *The Martian Chronicles*, as a way of repulsing a planetary invasion? "The Third Expedition" is as drenched with nostalgic happiness and pain as that excruciating final scene in Thornton Wilder's *Our Town* when Emily travels back in time to relive a few almost unbearable hours of her twelfth birthday.

There is Bradbury's tolerance—no, more than tolerance, his admiration—for intelligent nonconformity. In the country of the blind, Wells taught us in his finest short story, the one-eyed man is seldom king. "I have always looked upon myself as some sort of Martian," Bradbury once declared. Has any modern writer written more effectively against the book burners—that is, against the blind who tell us where and when to walk and what to read?

It is not just the political book burners that Bradbury despises. There is a curious type of moralist that one still meets today, in respectable intellectual circles, who actually believes that fantasy is unhealthy for children. I restrain myself from quoting from dusty-minded educators, librarians, psychologists, and juvenile-literature critics who have warned us of the baleful influence of Oz books on the young. "They filleted the bones of Glinda the Good and Ozma," Bradbury's William Stendahl phrases it, "and shattered Polychrome in a spectroscope and served Jack Pumpkinhead with meringue at the Biologists' Ball!"

It is one of those crazy attitudes that arouses Bradbury, as it did G. K. Chesterton, to a pitch of fury. How G. K. would have roared with laughter had he been privileged to read about the second fall of the House of Usher, that wildest of all the chronicles, in which Stendahl and Pikes carefully plot and carry out their fiendish revenge against a Moral Climates Investigator and all the members of the Society for the Prevention of Fantasy!

There is Bradbury's awareness of the enormous difficulty that human beings—and races and nations—have in trying to understand one another. With our minds surrounded by solid bone, the miracle of it is that, by using what Bradbury once called the "peek-holes" in our head, we are able to communicate at all. On rare and wonderful occasions, in Bradbury's fiction, two isolated, lonely souls actually do manage to touch and comprehend: for example, the love that develops between Bill Forrester (age 31) and Helen Loomis (age 90) in *Dandelion Wine* or the understanding between son and father that takes shape at the close of *Something Wicked This Way Comes*. But in the *Chronicles* it is a failure to communicate that occurs most often.

Sometimes the results are comic, like the partial communication by telephone between Walter Gripp and Genevieve Selsor. Sometimes the results are bitter, as when Southern whites, in the *Chronicles'* best-known episode, cannot comprehend why the town's blacks are so eager to build rockets and escape to Mars. Sometimes the results are sad, as when Tomás Gomez and Muhe Ca, momentarily thrown together by an inexplicable wrench of space-time, discover that each is a phantom to the other. They exchange a few bits of trivial information by telepathy but are as unable to clasp hands as you, dear reader, are unable to clasp the hand of a man or woman who lived in ancient Athens.

Sometimes the results are horrifying. The Second Expedition to Mars fails because the Martians, including one of their psychiatrists, cannot accept even the possibility of a race of people with pink skin instead of brown, blue eyes instead of yellow, ten fingers instead of twelve. In the country of the blind, the one-eyed man is mad.

I do not know whether Bradbury has read much of Chesterton, but *The Martian Chronicles,* like all of Bradbury's writing, glows with a Chestertonian mix of wonder, hilarity, exhilaration (and thankfulness?) at finding oneself miraculously alive in an endlessly fascinating universe. It cannot be said too often that Bradbury is not particularly interested in science. The scientific content of the *Chronicles*—what point is there in denying it?—is quite low. We learn very little about the actual Mars, and what we learn is, as we have seen, mostly wrong. We do learn a great deal about the colors and mysteries of *tellurian* experience. Going to Mars, like going anywhere, helps us take fresh looks at the too-familiar scenery of Green Town, Illinois. "Space travel," says Bradbury's unnamed philosopher in the book's epigraph, "has again made children of us all."

The descriptive touches in the *Chronicles* delight and startle the reader the way a rainbow, seen for the first time, startles and delights a child. Martian children play with toy spiders made of gold, spiders that spin filmy webs and scurry up their legs. Martian books are raised hieroglyphics on silver pages (aluminum in earlier printings of the stories) that speak and sing when fingertips brush over them. Martian airships are white canopies drawn by thousands of flame-birds. Blue-sailed sand-ships carry Martians over the sandy beds of dead seas, like the sand-ship that Johnny Doit built for

Dorothy and Shaggy Man to use in crossing the Deadly Desert that surrounds Oz. At night, Martians sleep suspended in a blue mist that in the morning lowers them gently to the floor. Martian guns shoot streams of deadly bees. The canals, cutting through mountains of moonstone and emerald, flow with green and lavender wine. Silver ringfish float on the rippling water, "undulating and closing like an iris, instantly, around food particles."

Bradbury is in love with the sights and sounds and smells of the world, and (like all true poets) he prefers to describe them with those simple, elemental words that are part of a child's vocabulary. Colors on Mars are red, blue, green, black, gold, silver—no fancy synonyms, just the old familiar color words. And the pages of the *Chronicles* are splattered with simple weather words: heat, cold, summer, winter, sun, stars, fire, ice, fog, rain, snow, wind.

Some year a college student will get a master's degree by counting and analyzing all those spots in Bradbury's fiction where the wind blows. One of his early stories is about a man obsessed by winds. Is wind a symbol of time and change? "And tonight—Tomás shoved a hand into the wind outside the truck—tonight you could almost *touch* Time." Does the wind remind Bradbury of happy boyhood days in Waukegan, Illinois, when he wore his tennis shoes and flew a kite? Is Bradbury contrasting the winds of Mars with the absence of wind in the stillness of interplanetary space? He may have missed on the canals, the oxygen, the orbital direction of a moon, but he scores a decided hit with his winds. They do indeed blow as furiously over the actual Martian sands as they do over the dream deserts of Tyrr.

The Martian Chronicles is, of course, the last great book that anyone will write about native life on Mars. But there are many reasons, some of which I have indicated, why the *Chronicles* will not cease to be the strange, beautiful, amusing, sad, and wise book it is. Critics have said it is Bradbury's best book because there is more science in it than in his other books. I believe the opposite to be true. *The Martian Chronicles,* I have argued, is as remote from science as *Something Wicked This Way Comes* and *Dandelion wine.* That is not a weakness of the book. It is one of its strengths. That is why the Mariner photographs failed to damage it. That is why, long after Mars has become a hearth to us, *The Martian Chronicles* will keep on stirring imaginations, arousing laughter and tears, and haunting the minds of those who have not forgotten how to read.

"You know what Mars is?" an old man at a filling station asks Tomás Gomez. "It's like a thing I got for Christmas seventy years ago—don't know if you ever had one—they called them kaleido-scopes, bits of crystal and cloth and beads and pretty junk. You held it up to the sunlight and looked in through at it, and it took your breath away. All the patterns! Well, that's Mars. Enjoy it. Don't ask it to be nothing else but what it is."

4   A Dreamer's Tales

The writings of Edward John Moreton Drax Plunkett, better known as the Irish baron Lord Dunsany, are enjoying something of a revival in this country—no doubt as part of a growing interest in fantasy, particularly among the young. Although Dunsany wrote traditional poetry, much of it not yet gathered in any book, and at one time had several plays successfully produced in Dublin and London, almost everyone agrees that his finest achievements were his early fantasy novels and short stories, especially the stories. In later years he turned more to the writing of humorous nonfantasy tales about a heavy-drinking prevaricator named Joseph Jorkens. They are skillfully plotted and worth reading, but it is to the earlier wonder tales that admirers of Dunsany return again and again—not for their plots or characters but for the strange beauty of their style, for their humor, and for the vivid inventions of Dunsany's imagination.

There is no other style quite like it. As a child Dunsany loved best the tales of Grimm and Andersen, but the major influence on his manner of writing, as he tells us in his autobiography, was the King James Bible. Its musical rhythms, especially in such passages as David's lament for Jonathan and the last chapters of Ecclesiastes, had for Dunsany what he called a "magical beauty" that moves the heart even when the words are only dimly understood. Dunsany's Biblical style is most evident in his early tales, of which *A Dreamer's Tales* (first published in 1910, when Dunsany was 32) was the fourth collection. You can open it almost at random and find lines that ring with Biblical overtones.

Let me lead slowly to one instance. In "Idle Days on the Yann," when Dunsany joins the ship's crew in prayer, he prays not to the

This article was originally the Foreword to *A Dreamer's Tales*, by Lord Dunsany (Owlswick Press, 1979), and is reprinted here with permission.

jealous God of the Old Testament but to one of the "frail affectionate gods whom the heathen love." The captain prays to the gods that bless fair Belzoond, his native city, while the sailors pray to gods of the smaller, neighboring towns of Durl and Duz.

Unlike J. R. R. Tolkien, who was a Roman Catholic, or James Branch Cabell, who was (or pretended to be) an Episcopalian, Dunsany never took the Biblical God—or any other god—seriously. Early in life, when he was enchanted by Greek mythology, he acquired a wistful pity for forsaken gods—gods once worshipped by millions but now remembered only in ancient myths. Perhaps it was this pity, Dunsany once speculated, that led him to create his own mythology—first the great gods of Pegāna (the resemblance to "pagan" is obvious, but the accent is on the long *a* as it is in "Dunsāny"), then hundreds of lesser deities. When Dunsany prays with the sailors he chooses the humble god Sheol Nugganoth, long deserted by jungle tribes—and, as far as I know, never remembered again even by Dunsany.

"Upon us praying," writes Dunsany, "the night came suddenly down, as it comes upon all men who pray at evening and upon all men who do not. . . . " The words remind a modern critic of E. E. Cummings's line, "The snow doesn't give a soft white damn whom it touches," but anyone familiar with the New Testament thinks at once of Jesus' remark (Matthew 5:45), ". . . for he maketh his sun to rise on the evil and on the good, and sendeth his rain on the just and on the unjust." Similar echoes of Biblical themes and rhythms chime through all of Dunsany's fantasies. His unearthly names for persons, gods, cities, geographical features—even flowers and musical instruments—resonate with the Near-East cadences of Biblical words. Even the Old Testament "begats" are here in "The Sword and the Idol."

In addition to his fondness for bizarre deities, Dunsany had a poet's longing for exotic realms that reminded William Butler Yeats (who edited the first anthology of Dunsany's plays and stories) of old Irish jewel work—"but more often still of a strange country or state of the soul that once for a few weeks I entered in deep sleep and after lost and have ever mourned and desired." But much as Dunsany loved the cities of his dreams, he loved nature more, especially its dawns and twilights, its great mists, and its glowing stars. For Dunsany all cities, even the mightiest, are doomed by the relentless passage of time to become once again the wild fields in which flowers grow and birds sing.

In "The Madness of Andelsprutz" we learn of a once proud but now conquered city that first went mad, then died like an ancient deity. Bethmoora is another dream city that expires, this time when a dread message is delivered. And what was the message? We are not told. Such withholding of essential information is one of Dunsany's favorite dodges. In this book we never learn why Carcassonne is such a splendid city. In *The Last Book of Wonder* Dunsany never reveals "Why the Milkman Shudders When He Perceives the Dawn," or why "The Bad Old Woman in Black" runs down the Street of the Ox Butchers.

Dunsany stoutly maintained that not once did he intentionally write an allegory. I am not sure this was entirely truthful, but in any case great fantasy always lends itself to allegorizing. Who can read "Poltarnees, Beholder of Ocean" without regarding the sea as a symbol of the vast unknown world that lies outside Plato's cave, beyond what Dunsany later loved to call "the fields we know"? Like all great fantasy writers Dunsany never confused these fields with those that lie beyond. This double vision is beautifully symbolized by the small, mysterious cottages on the banks of the Yann, so close to London that you can reach them (as we later learn in the second Yann story in *Tales of Three Hemispheres*) through the back door of a funny little shop on Go-By Street, just off the Strand. Windows on the west sides of these cottages look out on the world we know. Through their eastern windows you can see the "glittering elfin mountains, tipped with snow, going range on range into the region of Myth, and beyond it into the Kingdom of Fantasy. . . . "

Not only from nature and myths and imaginary cities but also from ugly commonplace things, Dunsany has the genius to extract mystery and wonder. Who could imagine a story ("Blagdaross") made of memories recalled by such discarded objects as a cork, an unburnt match, a broken kettle, an old cord that once strangled a suicide, and an abandoned rocking horse? Or a tale ("Where Tides Ebb and Flow") about a dead body buried for so many centuries in the mud of the Thames that its soul watches London slowly pass away and hears the singing birds return. And there are still stranger themes: a Hashish Man who discovers the secret of the universe but can remember nothing except that its Creator does not take the universe seriously, but sits in front of it and laughs. The mad occultist Aleister Crowley, by the way, wrote Dunsany a letter to praise this tale and to point out that Dunsany surely had never taken hashish (true) because the tale failed to confuse the normal ordering of things in time and space.

Dunsany's subtle humor, which becomes more prominent in his later books, can be found here in many places. When Dunsany parts from the friendly captain of the *Bird of the River*, with his priceless yellow wine, they shake hands—but "uncouthly" on the captain's part for it was not the custom in Belzoond. In "The Day of the Poll," a satire on those dreary elections in which one must choose between two political hacks, Dunsany persuades a man to abandon his vote by taking him to a field outside London. No matter. The election had been decided long in advance because one of the candidates had forgotten to contribute money to a football club.

Pervading *A Dreamer's Tales,* as it does all of Dunsany's wonder stories, are those simple, beautiful descriptive phrases that all of us who admire Dunsany find so hard to forget: "a night all white with stars," "the pure wild air that cities know not," "villages in valleys full of the music of bells," "the old wrinkled sea, smiling and murmuring song." And there are occasional sentences that startle: "Sometimes some monster of the river coughed."

When the soul of Poor Old Bill visits the Moon he finds it "colder there than ice at night; and there were horrible mountains making shadows; and and it was all as silent as miles of tombs; and Earth was shining up in the sky as big as the blade of a scythe, and we all got homesick for it, but could not speak nor cry." The passage is so scientifically accurate that one of our moon-walking astronauts might have written it had he been a poet.

"For how short a while a man speaks, and withal how vainly," Dunsany says in "The Idle City." "And for how long he is silent." Even a writer's recorded dreams last only a moment in the age of the universe, in turn but a moment in the endlessness of time. Yet, as long as the English language does not change too much, we can still hear the musical voice of Lord Dunsany as he remembers what he saw and did when he wandered eastward, beyond the fields we know, into the Land of Dreams.

5   The Computer as Scientist

If we could climb into a time machine and visit a research center of the far future, is it possible that we would find computers acting like theoretical scientists? Would they be not only recording data and telling robot technicians what to do but also making shrewd guesses about new laws and theories?

Let us fantasize. Robots fitted with sensitive devices for seeing, hearing, touching—perhaps even tasting and smelling—are performing complicated experiments suggested to them by a computer. The results of their sensory readings are transmitted to the computer, where they are systematically searched for patterns that confirm or falsify old conjectures and suggest new ones. On the basis of this analysis, more instructions are issued to the robots. The scientist has become an outsider. His job is to see that the lab operates smoothly and to make sure that its discoveries are promptly reported to other research centers around the world and translated into new technology.

Recent research on what are called computer "induction programs" suggests that this picture may not be as visionary as it seems. But first we must understand an important distinction—the difference between deductive and inductive thinking.

Deduction is the process by which statements in a formal system are obtained by logical inference from other statements in the system. It is entirely a matter of manipulating information or symbols according to prescribed rules. No observations of the outside world are required in deducing, for instance, that if all squirrels are rodents and if all rodents are mammals, then all squirrels are mammals.

This article originally appeared in *Discover*, June 1983, and is reprinted here, with a postscript, with permission. © 1983 by Discover Publications, Inc.

Induction requires looking at the world. It is the process by which scientists generalize from observations of individual instances to a universal law. If every electron measured is found to have one unit of electrical charge, it is assumed that all electrons, everywhere in the cosmos and at all times, have the same property. Induction is never absolutely certain. For all we know, there may be some vast cyclic law that will give every electron two units of charge next Tuesday. Nevertheless, evidence for a law may be so overwhelming that belief in its universality will come extremely close to certainty.

Both types of reasoning are used constantly in everyday life. If you assume that your car keys are either in your pocket or in the car and if you find that they are not in your pocket, you deduce that they are in the car. When you add up a restaurant check, you are applying a simple deductive algorithm (procedure) to arrive at the sum. But when you take a sip of wine, your expectation that it will not taste like coffee is an induction, a conjecture based on past experience.

We all know that computers are whizzes at deduction. Even your pocket calculator can deduce in a microsecond the product of two four-digit numbers. The deductive powers of a big computer extend far beyond number twiddling. Give it the posits and rules of any formal system, and it can make deductions with fantastic speed and efficiency.

Chess is a formal system. Computer programs can now deduce chess moves good enough to defeat even a grand master if there is a time limit of a few seconds per move. Some game-playing programs do a certain amount of induction, analyzing patterns and devising strategies on the basis of past experience. This is true also of the so-called "expert systems" that are now proliferating rapidly at centers for artificial-intelligence research. A computer is given information about a specialized field such as medical diagnosis and is programmed to deduce from Mr. Smith's symptoms that he has, say, the measles. Similar programs in geology can tell a company what the chances are of finding oil or copper at a specific spot. Although these systems use induction and give only probable conclusions, they are essentially deductive programs. They are little more than computerized textbooks, with rapid procedures for searching out desired information.

In recent years induction programs of a much more exciting sort have been developed. Like scientists, these programs search systematically through raw empirical data for regularities, then for-

mulate the simplest mathematical laws that can explain the regularities.

In an interview in 1983, Richard Feynman, a Nobel laureate
physicist at Caltech, likened the scientific method to the game of
guessing the rules of another game. Imagine, he said, that a man
who knows nothing about chess is allowed occasional glimpses of
small portions of boards on which chess games are in progress. It
does not take him long to realize that the board is an eight-by-eight
array of alternately colored squares. More observations lead him to
conclude that each player has one bishop that moves diagonally
only on dark squares and another bishop that is similarly confined
to white squares. Suddenly comes a surprise. He sees a game in
which a player has two bishops on black squares.

Have the rules of chess suddenly altered? No, because sooner
or later he observes the curious procedure by which a pawn that
reaches the opponent's first row may be exchanged for a bishop.
Other seeming violations of rules are explained when he sees such
rare phenomena as castling and capturing *en passant*. Because the
board, pieces, and rules of chess are finite, eventually he will obtain
a complete understanding of the game.

The game played by the universe is not so simple. Indeed, it
may be infinitely complex. Already, nuclear physicists are talking
about subquark entities that make up quarks that in turn make up
protons, neutrons, and other particles. Scientists are never certain
that what Einstein called the "secrets of the Old One" are completely understood. They may never be fully understood. Scientists
cannot even be sure that the rules will not change in time, as the
rules of Western chess have altered over the centuries. Nevertheless, nature seems to play fairly (it may be subtle, said Einstein, but
never malicious) and with a fixed strategy based on unalterable
laws. No one can deny that science has been fantastically successful
in learning some of these laws, especially after it discovered how to
extend observations by making complicated experiments and by
using ingenious instruments.

Why does induction work so well? How is it that human minds
can gaze at tiny patches of nature and formulate laws that have such
amazing powers to predict how all of nature will behave? Philosophers give different answers to such questions, arguing intermina-
bly over what they call the problem of "justifying" induction. Some
philosophers of science, such as Britain's Sir Karl Popper, have
abandoned the term induction altogether. But one way or another,
science must decide between competing conjectures, and no one,
not even Popper, doubts that science works. We know that comput-

ers are good at mathematics and chess and lots of other things, but can they be taught to play the science game? The answer is yes. The only debate now is over how well they can learn to play it.

A lot of fascinating computer research is under way on science induction programs. The basic scheme is simple. A computer is fed observational data about the outcome of experiments. It then searches this information for low-level equations that describe how the values of certain variables change with respect to one another. Suppose, for example, that a series of tests determines the intensity of light on a screen as the source of light is placed at different distances. A physicist would plot the results as spots on a graph; then, from the curve formed by the spots, he would guess the simple law: Light intensity varies inversely with the square of the distance. There are now computer programs that, given the same data, will quickly reach the same conclusion.

The most promising of recent programs of this sort are called BACON programs after Francis Bacon, one of the earliest philosophers to look for systematic procedures of inductive inference. These programs were developed in the late 1970s by Patrick Langley of Carnegie-Mellon University, one of the leading centers of artificial-intelligence research. Langley is now expanding and improving the programs along with Herbert Simon,who is best known for his pioneering work in artificial intelligence even though his Nobel Prize (awarded in 1978) was for economics.

The latest of these programs, BACON 4, is the work of Simon, Langley, and Gary Bradshaw, a graduate student at Carnegie-Mellon. When given data about the outcome of experiments, it has rediscovered scores of fundamental laws that were major discoveries in past centuries. It has formulated Archimedes' principle of floating bodies, Kepler's third law of planetary motion, Boyle's law of gases, Snell's law of light refraction, Black's law of specific heat, Ohm's law in electricity, and many others, including some basic laws of chemistry. And there are other induction programs at other research centers, such as meta-DENDRAL, developed by B. G. Buchanan and T. M. Mitchell at Stanford to generate and test hypotheses. There is even an induction program that discovers concepts in mathematics. It is called AM and was written by Douglas Lenat, also at Stanford. (By manipulating numbers and diagrams, mathematicians also experiment when they search for interesting theorems).

It is true that no induction program has yet found a new law and that the programs are not very good at filtering "noise" (errors) out of raw data. Critics of BACON contend that it works only on data that have been "cleaned" for analysis. Nevertheless, no one can see why, as induction programs improve, they will not some day be capable of discovering new laws. It was only a short time ago, remember, that skeptics of artificial intelligence were predicting that computers would never play chess above the tyro level.

Closely related to research on induction is work on programs that play induction games. What is an induction game? Consider the old parlor game that involves handing a pair of scissors around a circle of seated players. Each time the scissors are transferred, the person offering the scissors must say either "crossed" or "uncrossed." A moderator, who alone knows the secret rule, tells whether the player spoke correctly or incorrectly. The object of the game is to guess the rule. At first, most people suspect it has something to do with how the scissors are held, and experiments are made to test conjectures. Eventually the secret dawns on a perceptive player. One says "crossed" if one's legs are crossed and "uncrossed" if otherwise. Guessing the rule is an induction; the underlying regularity is obtained by generalizing from a set of observations.

A more sophisticated induction game, which models many aspects of scientific procedure, is Eleusis (after the site of the ancient Greek religious mysteries), a card game invented by Robert Abbott of New York City.

In Eleusis, players try to guess a secret rule that states what kind of card can be played on another. The rule is invented by the dealer, who, if the players prefer, can also be called God, Nature, Tao, Brahma, or the Oracle. Scoring in the game is cleverly designed so that it is to the dealer's advantage to think of a rule neither too hard nor too easy to guess. Too simple a rule would be: On every card play a card of opposite color. Of course the complexity of a rule likely to give the dealer a high score depends on how experienced the other players are. For beginners, a typical good rule would be: Play a black card on all cards with odd values, a red card on all with even values. How would you fare in playing Eleusis? As an exercise, study the sequence of played cards shown in figure 5.1 to see whether you can guess the simple rule that governs the sequence. Then check the answer, shown in figure 5.2 at the end of the chapter.

Many computer programs have been written for Eleusis, both to generate rules and to guess rules. It was not until the early 1980s, however, that workers in artificial intelligence created programs better at guessing Eleusis rules than were most human players.

Another induction game, Patterns, is based on visual patterns rather than sequential plays and was invented in the late 1960s by

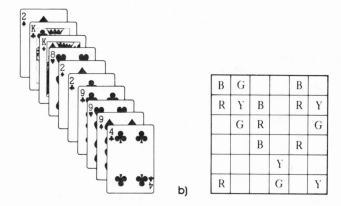

a)

b)

B	G			B	
R	Y	B		R	Y
	G	R			G
		B		R	
			Y		
R			G		Y

Figure 5.1: Problems in Inductive Thinking
a) Eleusis: These cards were dealt, from top to bottom, according to a rule. Can you discern and state the Eleusis rule?
b) Patterns: The lettered squares suggest a pattern that would emerge if the grid were filled in. Can you complete the grid?

Sidney Sackson of New York City. In Patterns, each player draws a six-by-six grid of squares. One player, called the Designer, secretly creates a pattern by coloring each square of his grid with one of four colors or by using four symbols—such as a square, triangle, circle, and cross and puts his sheet facedown on the table. As in Eleusis, scoring rules ensure that the Designer scores highest with a pattern that is neither too easy nor too hard for the other players to guess.

At any time a player may make an inquiry by putting a small check mark in the corner of one or more squares of his grid. The sheet is passed to the Designer, who must put in those squares the correct color (or symbol) of his secret pattern. Each square filled in by the Designer corresponds to the result of an observation of the facts by the players, who try to guess the pattern as soon as possible. The highest scores go to players who guess the most squares with the fewest inquiries. Low scores go to poor or unlucky guessers, to those who fill in squares too quickly (like scientists who rush into print with poorly confirmed conjectures), or to those so overcautious that they delay forming a hypothesis until others beat them to it.

Several computer programs play Patterns skillfully. Can you? To test your skill, see whether you can guess the pattern of the partly completed grid shown in figure 5.1, where the letters stand for colors red, yellow, blue, and green. (The answer appears at the end of the chapter.)

Programs such as BACON and programs for playing induction

games are similar to inductive programs that search cipher texts for patterns that may help break a code. There is a strong analogy between scientific induction and the kind of thinking that enables a person to solve a cryptogram or code. (Think, for example, of current work in cracking genetic codes.) And there are sophisticated induction programs now being used to analyze the noise received by radio telescopes, to see whether patterns can be found that would imply an extraterrestrial message. For all these reasons, there is growing confidence that computers some day may indeed be able to discover new scientific laws.

As for scientific *theories,* that is a different ball game. Although no sharp lines divide laws from theories, just as no sharp lines separate facts from laws, the distinctions are obvious and useful. Laws are descriptions, usually mathematical, of how observable quantities are related. The word "observable" is extremely fuzzy, but it usually means a property of nature that can be observed and measured in simple, direct ways: length, volume, mass, velocity, momentum, color, pitch, and so on. Properties observed through special instruments such as microscopes and telescopes are counted as observables because the process is so simple that no one doubts that what they see is actually there.

Theories involve "unobservable" concepts such as electrons, neutrinos, quarks, electromagnetic fields, gluons, and a thousand other ghostly things. Can you imagine anything less observable than a gravity field or the wave function of an atom? Yet relativity and quantum mechanics could not do without these concepts. Laws are designed to account for facts. Theories are constructed to explain both laws and facts. There are algorithms for getting laws from facts (otherwise the BACON programs would not work), but there are no known algorithms for getting good theories from facts and laws.

Are algorithms for devising good theories possible in principle, or do they require some mysterious creative ability of the human mind that will be forever beyond the reach of computers? Of course, once a theory is constructed, empirical consequences can usually be deduced, and then observations can confirm or refute the claims. General relativity, for example, with its incredible assertion that gravity and inertia are the same force, was poorly confirmed until recently. Now it is supported by hundreds of tests made possible by atomic clocks. How the hardware and software inside Einstein's skull arrived at the theory remains a mystery.

Philosophers argue fiercely over whether the intuitive leap needed for theory construction can be simulated by a computer. Some think that artificial intelligence will never formulate a good theory. Simon is among the optimists. In his view, theory invention is simply a more complicated level of problem solving, and he points out that in the process of analyzing data and searching for laws, his BACON programs actually introduce new low-level theoretical concepts. Like most of his colleagues in artificial intelligence, Simon believes that there are no good reasons for doubting that computers will eventually be programmed to do any kind of thinking a human mind can do.

We are, of course, now deep into metaphysical questions about the nature of human consciousness and creativity. As to whether Simon is right or not, we shall just have to wait and see—or perhaps wait and not see.

Answers

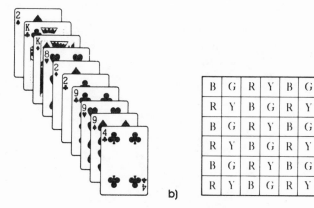

B	G	R	Y	B	G
R	Y	B	G	R	Y
B	G	R	Y	B	G
R	Y	B	G	R	Y
B	G	R	Y	B	G
R	Y	B	G	R	Y

a) b)

Figure 5.2: Answers to Problems in Inductive Thinking

a) Eleusis: The Eleusis rule is to deal a card that matches the previous card either in color or in value.

b) Patterns: The completed board above shows one of the possible patterns that can be induced.

Postscript

For detailed rules on how to play Eleusis, see both the chapter on this game in my *Second Scientific American Book of Mathematical Puzzles and Diversions* (1959) and my follow-up column on improved versions of the game in *Scientific American,* October 1977. For the rules of Patterns, see both Sidney Sackson's book *A Gamut of Games* (1969) and the chapter on Patterns in my *Mathematical Circus* (1979).

In 1984 the Fredkin Foundation, established by Edward Fredkin, an artificial-intelligence expert at MIT, announced a prize of $100,000 for the first computer program to make a mathematical discovery. More precisely, the discovery must be a major new theorem based on mathematical ideas not implicit in the program that discovers it. A committee of distinguished mathematicians will rule on the award.

The foundation has an award of the same amount for the first computer program to become the world's chess champion in tournament play with human grand masters. As of now, no one seems even close to winning either prize.

Illusions of the Third Dimension

The complicated process by which animals and human beings see and interpret the outside world has long been an exciting research area. An important aspect of that research is the study of how flat surfaces can be made to simulate the illusion of seeing a real world. Paintings and photographs do a fairly good job of tricking the brain into thinking it is looking at an actual scene. The stereoscope adds an illusion of depth. Motion pictures strengthen the illusion still more by introducing action and sound. The next big step is to make motion pictures three-dimensional (3D) without the need for those annoying cardboard glasses.

Sooner or later movies and TV will surely go 3D, but exactly when or by what means remains uncertain. Today, in research centers around the world, psychologists are collaborating with physicists on a variety of strange systems for seeing films and videotapes in depth, but, before examining some of them, we must first understand how psychologists explain the process of seeing three-dimensionally.

Cup your hand over one eye. At first there seems to be little change. Your brain is inferring depth from such cues as perspective, size, overlapping, variations in color and texture and shading, haze on distant scenery, and so on. Motions are important. When you move your head, close objects shift more rapidly across your visual field than do distant ones. When one drives along a road, nearby trees whiz by but clouds seem to follow the car. Moreover, the lens of each eye adjusts its shape to the distance of an object. Look with one eye at a finger near your nose—distant objects blur. Look far away—the finger is out of focus. This, too, adds depth to monocular vision.

This article originally appeared in *Psychology Today,* August 1983, and is reprinted here, with changes and a postscript, with permission.

If, however, you keep one eye covered for several minutes and then uncover it, the increase in depth perception is startling. The reason, of course, is that each eye views the world from a different angle. Your left eye sees more of the left side of an object, and your right eye more of the right side. The closer an object, the more you see around it. Your brain fuses the two disparate images to give a strong sensation of depth. It is this binocular illusion, called stereopsis, that must be simulated if movies and TV are ever to become truly three-dimensional.

How do your eyes convey information to the brain? The process is utterly fantastic. Until Newton suggested otherwise, it was believed that each optic nerve went to the same side of the brain. Now we know that the nerves cross—but in a manner so bizarre that biologists are still puzzling over why it evolved. All nerve fibers from the left side of each retina go to the left brain, all from the right side to the right brain. It is crazier than that. Because the eyes' lenses turn images upside down on each retina, the entire left side of your visual field goes to the right brain, the entire right side to the left brain. A totally invisible seam runs vertically down the middle of your visual field. Your right brain "sees" everything left of this line, your left brain "sees" everything on the right!

How the brain fuses the two streams of impulses to create a solid, seamless world, "out there," remains a total mystery. It was once thought that the millions of optic fibers go to single regions on each side of the brain to create tiny "maps" of the world. But no—the fibers lead to widely scattered regions in the midbrain and the visual cortex. There are no maps. There is only an incredibly complicated process of coded information transmittal and interpretation that nobody understands.

The first attempt to simulate binocular vision was by using the stereoscope. It was invented in 1833 by Sir Charles Wheatstone, an English physicist. When two mirrors were placed at right angles, each eye could separately see the two pictures on opposite sides of the device. If the pictures represent what each eye would normally see, the brain merges them and strong stereopsis results.

In 1978, when Manhattan's Guggenheim Museum gave the world's first public exhibition of a stereoptical painting, it was viewed in just this way. Salvador Dali had produced two nearly identical pictures entitled "Dali Lifting the Skin of the Mediterranean to Show Gala the Birth of Venus." Visitors walked toward two large mirrors placed at a 60-degree angle. The reflections of the two paintings slowly moved together to become a 3D picture when a visitor's nose almost touched the corner of the mirrors. The optical system had been devised by Dali's Manhattan friend Roger de Montebello, an expert on stereopsis who has his own patented ways of providing wide-angle 3D photos. Dali has made many stereoscopic paintings since Montebello showed him how to do them.

Wheatstone discovered that when he switched the pictures in his stereoscope there were strange depth reversals. This led to his invention of the "pseudoscope" for viewing the world through prisms that exchanged the eyes' visual fields. The result was magical. Spheres looked concave. Hollow objects became convex. A marble rolling inside a bowl seemed to roll around a hill until it came to rest on top.

Do faces turn inside out when seen through a pseudoscope? They do not. Seeing is a process by which the brain unconsciously forms hypotheses about the world and then rapidly selects the best bet in the light of one's total experience and (perhaps) genetic inheritance. Because we never see people with faces that resemble the insides of masks, our mind is incapable of making the nose go backward even when it is seen through a pseudoscope. A statue of a person inside a wall niche remains normal in a pseudoscope, even though the niche reverses and the statue seems to project from the wall! On the other hand, the back of a face mask looks like a normal face in a pseudoscope. Indeed, it is so easy for the mind to reverse a concave face that when backs of masks are viewed at a distance it is difficult *not* to see them as convex.

I wish now that I had bought a beautifully painted inside-out marble bust of Jesus I saw on sale many years ago in an antiques shop. As you walked past, the head seemed to rotate so that the eyes followed you. Try hanging on the wall either (1) a rubber head mask that has been turned inside out and cut in half so you can view the painted but now concave side or (2) a plastic mask turned around. It is best to have the mask tilted slightly backward, high on a wall, and illuminated from below. Close one eye and move from side to side. The face will appear normal and seem to turn as you move.

The mind's tendency to interpret flat pictures in the light of experience is the basis of many astonishing depth illusions. Photographs of the moon's craters are a familiar example. We are so used to seeing things illuminated from above that when you turn photographs of the moon's surface the "wrong" way, craters instantly became flat-topped mounds. Where is the missing slice of pie in figure 6.1? Invert the picture to find out! Because we never see pie attached to the underside of plates, our mind makes the most plausible guess about what the picture represents in each orientation.

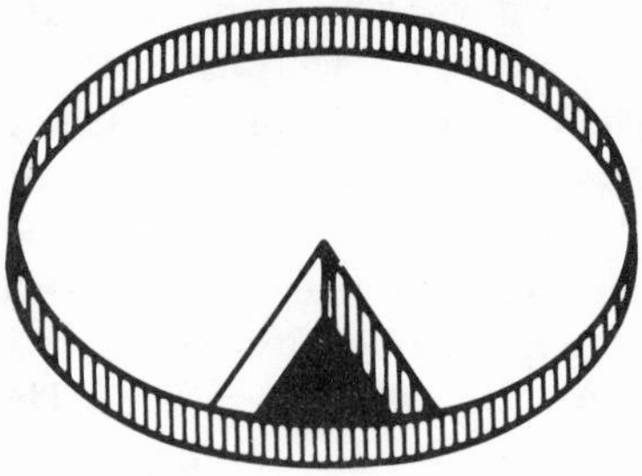

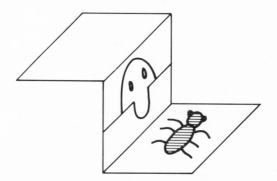

Figure 6.1: Find the Missing Slice

Figure 6.2: Can Kilroy See the Bug?

When experience is no help in interpreting an ambiguous drawing, the mind oscillates between alternate hypotheses. In figure 6.2, can Kilroy see the bug?

Many researchers have been experimenting in recent years with the brain's ability to reverse the appearance of familiar convex objects when the latter are modeled as concave structures. The El Paso Science Center in Texas gives away a cardboard illusion of this sort designed in 1980 by Fred and Ellen Duncan. You cut and fold it to make three sides of a large die with black spots on the *insides* of the faces. If you hold it so the concave side faces you and then stare with one eye at the corner where the spotted faces meet, the die will soon snap inside out in your mind and appear normal. When you move your hand, the die seems to turn the opposite way. A similar card designed by magician Jerry Andrus—he calls it a "parabox"—folds to resemble a wooden crate. (The parabox and other remarkable illusions can be obtained from Andrus by writing to him at 1638 East First Avenue, Albany, OR 97391.)

Duncan, Andrus, and others have discovered that an unusually strong version of this illusion is obtained with a model of an inside-out house. Photocopy the house in figure 6.3 (based on a model by Andrus), mount it on thin cardboard, then cut it out. Cut line *AB*. "Valley fold" lines *BC, BD,* and *BE*. Paste the triangular flap to the back of the roof so that the front and side of the house are at right angles.

Put the concave house on a shelf at exactly eye level. Stand about ten feet away and observe the model with one eye. As soon as it snaps to normal, walk to one side (keeping one eye closed); you will see the house rotate the wrong way. After you are practiced at "locking in" the reversed perception, let the model rest on its back, on your palm, and look into it with one eye. Tilt you hand various ways. The conflict between what you see and feel is indescribable. What happens if you cut two sides of the door and open it a trifle outward? Or if you push a pencil halfway through one of the windows?

Wheatstone could use only drawings for his first stereoscope, but photography soon provided better pictures. David Brewster, another English physicist, improved the instrument by using lenses instead of mirrors, and by 1860 scarcely a parlor in France or England was without a Brewster stereoscope. It was Oliver Wendell Holmes who designed the familiar U.S. model. Brewster's clumsy box was replaced by a lightweight hood to go around the eyes, a handle for one-hand holding, and a trombone slide for focusing. Today, color transparencies are viewed stereoptically through small plastic devices, but the old stereo cards are avidly collected.

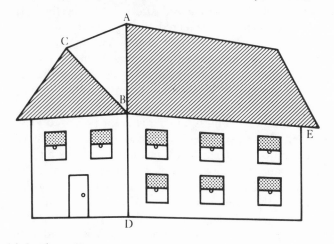

Figure 6.3: Inside-out House

When motion pictures began, it was at once apparent that they could be made stereoptic. By the end of the silent-film era there were hundreds of patents for such systems, some with the two images side by side on the screen, others with one image above the other. All were impractical because they required cumbersome viewing devices to be either worn on the head or mounted in front of each seat in a theater.

One strange system—it generated scores of patents—alternated the frames of left and right scenes on the screen. To allow each eye to see only the proper frame, viewers looked through noisy devices or spectacles that contained rotating disks or oscillating shutters that alternately blocked each eye in synchronization with the projector. A few years ago shutter spectacles that produce only a faint hum were designed to operate by the piezoelectric effect rather than mechanically.

The first practical way to eliminate expensive viewing equipment was the application of an 1850s discovery that used complementary color filters. The pictures for one eye are projected through, say, a red filter, the others through a green. The two colors overlap on the same screen, but, when viewed through the now familiar red-green spectacles, each color is eliminated for one eye, and the brain fuses the red and green images to produce a gray picture with depth.

The first 3D motion picture to use the "anaglyphic" method, as it was called, was a 1922 film, *The Power of Love*. A few other crude anaglyphics, producing lots of eyestrain, were made in the thirties and forties. The system was soon replaced by a much superior one based on the polarized filters developed by E. H. Land. Right and left films are projected through polarizing filters with their axes of polarization at right angles. Viewing spectacles have polarized filters similarly oriented so that each filter blocks light from one of the projected films, allowing light from the other film to reach the other eye. The system permits stereoptic viewing in full color.

The first feature-length movie made for polarized viewing was *Bwana Devil* (1952), starring Robert Stack. The next few years saw the production of more than fifty polarized features and almost as many shorts and cartoons. The best money-makers were horror clinkers such as *The House of Wax* and *Creature from the Black Lagoon*. Many of these films, such as Alfred Hitchcock's *Dial M for Murder*, were also released for normal viewing. (Audiences seeing them flat were puzzled by how often objects got tossed at the camera.) The polarized features enjoyed a brief vogue, along with a rash of anaglyphic comic books, before the novelty wore off. You will find a complete listing of 3D motion pictures and comic books in *Amazing 3-D* (1982), by Hal Morgan and John Symmes, along with orange-blue glasses for viewing the book's many illustrations. In recent years some X-rated and porno flicks have been made for polarized viewing.

Polarized 3D cannot, of course, be transmitted by TV, but James Butterfield has developed a process that converts old polarized movies to videotape for anaglyphic viewing. In 1980 Rita Hayworth's *Miss Sadie Thompson* was shown for anaglyphic viewing on cable TV in selected U.S. cities, and more recently other polarized films have been similarly adapted for television.

The most peculiar method of simulating depth on screens, though only for images in horizontal motion, is based on a little-known illusion called the Pulfrich pendulum. You can demonstrate it easily. Tie an object to one end of a piece of string. Have someone swing the bob while you observe it with a dark glass over one eye. (A sunglass lens will do, or colored cellophane, even a card with a pinhole.) Keep both eyes open and look at the background, not at the bob. Unless you have a strongly dominant eye, you will see the bob trace an ellipse! Switch the dark glass to the other eye; the bob will change its direction of revolution.

Let us call sunglasses, with one lens removed, Pulfrich spectacles. Wear them while a passenger in a moving car and you will find that the car's speed seems different, depending on which side you look. On the side where speed seems slower, houses and trees appear larger than normal. On the other side they seem dwarfed. Stand on the sidewalk and see how the traffic lanes bend. When TV "snow" is viewed through Pulfrich glasses, you see two layers of dots that drift opposite ways.

The Pulfrich effect was discovered about 1920 by Carl Pulfrich, a German physicist who could not see the illusion because he had lost an eye in an accident. On the basis of data supplied by associates, he wrote a paper in 1922 in which he correctly guessed what is now recognized as the cause of the phenomenon. Darkened images on the retina take a microsecond longer to reach the brain than do bright images. Consequently, the eye behind the dark glass sees a moving object at a position slightly in the past relative to the image on the other retina. If the motion is horizontal, the brain interprets the fused images stereoptically.

You may have seen advertisements in the mid-sixties for a miraculous pair of glasses said to add 3D to movies and TV. What you got for $9.95 was a cheap pair of Pulfrich sunglasses. For many years

Tokyo TV has been showing animated cartoons for children who watch through Pulfrich spectacles. The plots are cleverly planned so that there is lots of horizontal action to create depth appropriate for the story line.

Many other 3D systems are being developed by companies hopeful of commercial success. The DOTS system (an acronym for digital optical technology systems) puts near and far objects slightly out of focus, with red-green color fringes, although the picture itself is in natural color. The film is observed through lightly tinted red-green glasses.

An unusual system, called Visidep, is being developed by CJM Associates, at Chapin, South Carolina. The letters stand for the three inventors, physicists LeConte Cathey and Edwin R. Jones and media-arts specialist Porter McLaurin, all professors at the University of South Carolina in Columbia. Their technique requires no modifications of screen projection equipment or TV sets and can be viewed without any visual aids. Indeed, the depth illusion is just as strong for a person with one eye!

Visidep is based on the fact that when you move your head, close-up objects shift more than distant ones. An ingenious camera is designed to create on the film or videotape a slight flutter of images, the flutter being greatest for close-up objects and steadily diminishing with distance. The background remains unmoving. The flutter is distracting, but the inventors have a method, which they are not disclosing until patents are obtained, that will diminish the flutter. The simulation of depth is remarkable, even though it is not produced binocularly. It adapts beautifully to videotape and computer graphics, for displaying models of molecules and other solid geometrical structures. Videotapes based on the system were first shown to the public on the ABC Evening News in August, 1982.

A much older system of "naked viewing," based on genuine stereopsis, is used in those three-dimensional advertising displays of ladies who wink as you walk by. The system is also the basis for 3D postcards, greeting cards, and wall pictures. There are even ways to print such pictures in magazines. (The first was in *Look,* February 25, 1964.) Techniques vary, but the essential idea is to slice two or more pictures into vertical strips that can be hairline thin, interlace them side by side, and cover them with a plastic coating of vertical ridges that act as cylindrical lenses. If the head is held straight, the left eye sees only the strips forming the picture for the left eye, and similarly for the right eye. If more than two pictures are serrated, you can move your head from side to side and actually see slightly around close-up objects.

These "lenticular" sheets, as they are called, are simplifications of the much older "integral" system announced in 1908 by its inventor Gabriel Lippmann, a French physicist and Nobel Prize recipient. A "fly's eye" sheet is covered with tiny spherical convex "lenslets," each on top of a complete miniature photograph of the scene. Roger de Montebello's technique (mentioned earlier) is a great improvement over Lippmann's. It provides true depth and allows tilting the sheet in all directions without any "jumping back" of the image. The lenslets, in hexagonal array, permit a full-color 3D image that can be "looked around" from any angle within 90 degrees.

The Nimslo camera, named for its promoter Jerry Nims and his Chinese inventor friend Allen Lo, uses a lenticular (ridges) system. The instamatic-type camera takes four side-by-side pictures with 35-mm color film, which must be sent to the firm's headquarters in Atlanta for developing. Timex has invested $100 million in the camera, now on sale throughout the United States. Its pictures tend to resemble cardboard cutouts, but this may be remedied by a new camera that will take six pictures simultaneously.

Motion picture screens with lenticular grids have been operating in several Soviet theaters since 1941, when the first trial film, *Concerto,* was shown in Moscow. It was soon followed by other feature-length films, some of which have been released for flat viewing. Engineers in Russia and elsewhere are struggling with ways to apply similar techniques to TV—and to develop quite different techniques, some top secret. A novel method of displaying computer graphics in true 3D is now commercially available in the United States under the name SpaceGraph. It uses a flexible mirror that vibrates rapidly, projecting images at different depths on a series of planes. From any viewing angle the images on the computer screen seem to float in space, and no special glasses are required. (See *Discover,* May 1982.) Unfortunately, this and other naked viewing systems are much too complex and costly for motion picture or TV use.

The ultimate in 3D realism, as everybody knows, is holography. A holographic image, created without lenses by using laser beams, is indistinguishable from what you would see if you looked at the actual scene through a window. Holography may provide the system that will eventually become standard for motion pictures and TV, but the technical problems to be overcome are enormous. Trying to guess when it will be commercially profitable is like trying to guess when solar energy will replace fossil fuels. But as the great Russian film director Sergei Eisenstein once wrote, "It is as naive to doubt that the stereoscopic film is the tomorrow of the cinema as it is is to doubt that tomorrow will come."

Postscript

The Visidep system originally introduced parallax displacement by jiggling the camera horizontally. In a paper published in 1984, Jones, McLaurin, and Cathey reported their surprising discovery that the depth illusion is enhanced if this jiggling is vertical. They conjecture that this happens because "we bounce vertically when we walk, but we are not conscious of the resulting motion in our visual field. Also in normal vision our eyes converge only in the horizontal direction and track identically in the vertical direction. Thus for vertical motions we are used to receiving identical parallax information in both eyes."

Artist Terry Pope, a lecturer in the Department of Fine Art, University of Reading, England, makes and sells what he calls Phantascope 2, a pseudoscope that exchanges right and left visual fields by a clever mirror arrangement. He also makes several kinds of rotating structures, parts of which seem to turn in opposite directions when viewed through the device. His Phantascope 1 is a "hyperscope" whose mirrors cause you to see as if your eyes were eight inches apart. If a person puts the back of a hand on his nose, through the hyperscope it looks as if the hand is two feet in front of his nose. Through the pseudoscope a tree A, nearer to you than tree B, seems to be farther from you than B. (See Jearl Walker's "Amateur Scientist" department in *Scientific American*, November 1986, for a description of Pope's optical devices.)

The final episode of the TV series "Moonlighting," scheduled for May 1988, was supposed to carry an episode designed to be viewed through cardboard spectacles that would be distributed around the nation by Coca-Cola, the film's sponsor. Because of a writer's strike, this had to be postponed. I am told that it will take place in the fall of 1988. The glasses, with one darkened lens, will rely on the old Pulfrich effect to create depth illusions for specially designed horizontal motions. Coca-Cola's publicity releases called Terry Beard the inventor of the process (see *Newsweek*, February 15, 1988), though exactly what he invented is not clear.

Newsweek also reported that Toshiba will be introducing in 1988 a $2,800 camera for taking 3D movies to be viewed through glasses that contain liquid-crystal shutters. The glasses are wired to a screen that rapidly alternates right- and left-eye images. The shutters are synchronized so that the left eye sees only the left image and the right eye sees only the right image. Similar systems have been announced by other Japanese firms, but of course none is applicable to TV broadcasting.

Person and Place Names

Many years ago, in one of my *Scientific American* columns, I had my numerologist Dr. Irving Joshua Matrix posing as a fake psychiatrist. Matrix explained how the names of people often play a strong unconscious role in shaping their character and life history. I thought I was inventing something new, but it turns out that Carl Jung was way ahead of me. Here is a footnote from page 11 of the Bollingen paperback edition of his book *Synchronicity: An Acausal Connecting Principle:*

> We find ourselves in something of a quandary when it comes to making up our minds about the phenomenon which Stekel calls the "compulsion of the name." . . . For instance, Herr Gross (Mr. Grand) suffers from delusions of grandeur, Herr Kleiner (Mr. Small) has an inferiority complex . . . Herr Feist (Mr. Stout) is the Food Minister, Herr Rosstauscher (Mr. Horsetrader) is a lawyer . . . Herr Freud (joy) champions the pleasure-principle, Herr Adler (eagle) the will-to-power, Herr Jung (young) the idea of rebirth, and so on. Are these the whimsicalities of chance, or the suggestive effects of a name, as Stekel seems to suggest, or are they "meaningful coincidences"?

It seems to me that Herr Jung's name more obviously symbolizes the "young" enemy of Father Freud. As for the name of psychoanalyst Herr Stekel (Mr. Little-stick), Jung surely overlooked another obvious bit of Freudian symbolism.

In 1966 Richard Nixon complained that the DuBois Club for young blacks had used Communist deception in choosing its name. The club had been named for the black sociologist W. E. DuBois,

This article originally appeared in *Word Ways,* November 1983, and is reprinted here, with a postscript, with permission.

who joined the Communist Party at age ninety-three and died an expatriate in Ghana. DuBois pronounced his name "DooBoys" rather than in the French manner; hence the club's name sounded exactly like "The Boys Club" of which Nixon was then national board chairman. The reader is referred to *The New Yorker's* "Talk of the Town," March 19, 1966, for the amusing details.

Authors often anagram their names to get pseudonyms. It is not generally known that Alexander Graham Bell adopted a pen name because he suspected that his articles were being accepted by magazines only because of his fame. Wanting them accepted on their merit, he sold several articles to *The National Geographic,* submitting them under the name of H. A. Largelamb, an anagram of A. Graham Bell. The articles all appeared under Largelamb's byline.

William Remme of Eureka, California, has called my attention to two oddities involving Ronald Wilson Reagan. Not only does each name have six letters, yielding the Biblical number of the Beast, 666, but if you add the values of the letters (using the cipher $A = 100, B = 101, C = 102$, and so on) you get a sum of 1984. (In 1983 I took this to be a certain prediction that Reagan either would or would not be reelected president in 1984.)

The *Newsweek* Feature Service in September 1971 distributed an interesting release by Edward Blau on American town names. Blau disclosed that Peculiar, Missouri, was named by a postmaster who had been asked to think of a name peculiar to his area. The founders of Odd, West Virginia, so named it because they wanted an odd name. Extra Dry Creek, Arkansas, was called that because it is even drier than nearby Dry Creek. Wynot, Nebraska, was named because no one could see why not. Blau recommended George R. Stewart's *American Place Names* as the best source for information on such oddities.

Unintended Puns

The following sentence from the third edition of *Principles of Mechanics* by John L. Synge and R. A. Griffith was sent to me by Phillip Morgan: "*Space* does not permit us to attempt an axiomatic treatment of the theory of relativity."

Harold Bloom, reviewing Norman Mailer's recent monstrosity *Ancient Evenings*—the greatest literary experience for me in 1983 was not reading this novel—for the *New York Review of Books,* April 28, 1983 (Page 4), had this to say: "In *Ancient Evenings* he has emancipated himself, and seems to be verging on a new metaphysic, in which heterosexual buggery might be the true *norm.*"

In the chapter on aesthetics in my *Whys of a Philosophical Scrivener,* I give two classic instances of unintended puns by famous poets (involving the words *raspberry* and *balls*), but I failed to mention the most incredible instance of all. Near the end of the last part of *Pippa Passes,* Browning writes:

> Sing to the bats' sleek sisterhoods
> Full compliance with gallantry:
> Then owls and bats,
> Cowls and twats,
> Monks and nuns, in a cloister's moods,
> Adjourn to the oak-stump pantry!

It is hard to believe, but Browning was so unfamiliar with street slang that, when he encountered the word *twat* in an old book of rhymes called *Vanity of Vanities* he assumed it referred to part of a nun's attire that he could appropriately pair with the cowls of monks! (See the entry on *twat* in *The Century Dictionary.*) Even more incredible is the fact that Browning never altered the lines. It is possible no one told him? I would welcome hearing from any Browning expert who could provide more details about this memorable literary gaffe.

Intended Dirty Word Play

As all students of Shakespeare know, the bard was fond of off-color linguistic jokes, the bawdiest of which are not likely to be footnoted even in scholarly editions of Shakespeare's works. Surely the most outrageous example occurs in act II, scene V of *Twelfth Night.* Malvolio is reading a letter from Olivia:

> By my life, this is my lady's hand: these be her very C's
> her U's, and her T's; and thus makes she her great P's.

Observe how the word *and* supplies the *N,* and how the letter *P* continues the joke. No wonder Lewis Carroll thought that Thomas Bowdler's edition of Shakespeare should be further censored. "I have a dream of Bowdlerising Bowdler," was how he put it in a letter, "i.e. of editing a Shakespeare which shall be absolutely fit for *girls.*"

Terse Verse and Short Fiction

In a footnote to C. C. Bombaugh's *Oddities and Curiosities of Words and Literature,* which I edited for Dover, I discussed some of the famous short poems in English. Others include such oldies as "Hired. Tired? Fired!"; "The Bronx? No thonx!"; and "Candy is dandy, but liquor is quicker," to which Ogden Nash added the line, "Pot is not." This subject was also explored in the *November 1981 Kickshaws,* where Jeff Grant quoted a number of three-word poems by Samuel Beckoff.

When William Cole wrote an essay on "One-line Poems and Longer, But Not Much" (*New York Times Book Review,* December 2, 1973), the review later published (January 13, 1974) a letter from G. Howard Poteet in which he proposed one-letter poems: "Thus my work includes the most evocative of all poetic letters, *O.* Further, there is the egocentric poem, *I,* the poem of pleasure, *M,* the scatalogical verse, *P,* the somnabulistic bit of poesy adapted from the comics, *Z.*"

I suggest we take this a step further with the following poem titled "Simplicity":   . No one can say my poem does not have a point. Of course we can write an even simpler poem, completely pointless, with the title "Ultimate Simplicity." It goes like this:

For many years there were efforts in American science fiction magazines to write short-short-short-short stories. One of the best was titled "The Shortest Horror Story Ever Written."

> The last man on Earth sat alone in a room. There was a knock on the door.

Ron Smith shortened this one letter by changing knock to lock.

Forrest J. Ackerman holds the record for brevity. In the 1970s he sold the following story to *Vertex* for $100:

Cosmic Report Card: Earth

F

At a 1983 science fiction convention which I attended, Ackerman said he has since resold his story four times for the same amount and that it has been translated into three languages. In case anyone tries to imitate it with other letters, he added, he has all twenty-six copyrighted.

Great Doggerel

Readers of my *Whys of a Philosophical Scrivener* will know of my fondness for poetry so terrible that it is funny. Some of the worst poetry ever published was written by the famous British-American mathematician J. J. Sylvester. As *The Dictionary of American Biography* delicately puts it, "most of Sylvester's original verse showed more ingenuity than poetic feeling." His privately printed book, *Spring's Debut: A Town Idyll,* is a poem of 113 lines, every line ending with the sound "in". Another long poem, *Rosalind,* has about 400 lines, all rhyming with "Rosalind". Here is how Sylvester's successor at Johns Hopkins University described an occasion on which Sylvester recited his poem to a meeting of the Peabody Institute:

> The audience quite filled the hall, and expected to find much interest or amusement in listening to this unique experiment in verse. But Professor Sylvester had found it necessary to write a large number of explanatory footnotes, and he announced that in order not to interrupt the poem he would read the footnotes in a body first. Nearly every footnote suggested some additional extempore remark, and the reader was so interested in each one that he was not in the least aware of the flight of time, or of the amusement of the audience. When he had dispatched the last of the notes, he looked up at the clock, and was horrified to find that he had kept the audience an hour and a half before beginning to read the poem they had come to hear. The astonishment on his face was answered by a burst of good-humored laughter from the audience; and then, after begging all his hearers to feel at perfect liberty to leave if they had engagements, he read the Rosalind poem.

Sylvester explained his idiosyncratic views on poetic structure in a little book called *The Laws of Verse,* published in 1870.

When I was a high school student in Tulsa, an English teacher asked everybody in the class to write a poem. A friend who sat next to me produced a poem that I thought such a masterpiece that I have carefully preserved it over the decades. Here it is, word for word, exactly as he wrote it:

Great Smells

A smell is the greatest joy seen.
  A smell that makes a new world serene.
Of all the smells of my pickin'
  I believe I would rather smell chicken.

The smell of chicken is very fine.
The smell of chicken makes me feel divine.
There are smells of cake and pie,
But the smell of chicken is enjoyed by I.

I smell the smell of aroma of coffee,
I smell the smell of the deep blue sea;
The smell of melon and the smell of meat,
But the smell of chicken can't be beat.

Mnemonics

Leigh Mercer, the London wordplay expert who wrote "A Man, A Plan, A Canal—Panama!" and other fine palindromes, sent me the following bewildering paragraph which he had clipped from a newspaper. The author was one F. E. White: "If you remember how much easier it is to remember what you would rather forget than remember, than remember what you would rather remember than forget, then you can't forget how much more easy it is to forget what you would rather remember than forget what you would rather forget than remember."

I once mastered an ingenious mnemonic system for remembering words and numbers, but I long ago forgot it. One of the country's top experts on mnemonics is the magician Harry Lorayne. Perhaps you have seen him perform his great memory act in person or on television. A magician friend recently told me that he used to forget names but that his memory enormously improved after he read a book on mnemonics by Harvey Lorayne.

Acronyms

In his autobiography (P. 150), Gilbert Chesterton tells how he and his friends once formed a club in London that they called IDK. Whenever anyone asked what the letters stood for, the reply was always "I don't know." I'm sure many readers of *Word Ways* have seen the sign WYBADIITY that hangs in bars. If a customer asks what it means, the bartender replies: "Will you buy another drink if I tell you?"

Vladimir Nabokov, in his novel *Pnin,* introduces the phrase *motuweth frisas.* Clearly it refers to the six days following Sunday.

Here are some useful acronyms for the most often repeated phrases in speeches by American politicians: *Bomfog* (brotherhood of man under the fatherhood of God), *Fisteg* (fiscal integrity), *Moat* (mainstream of American thought), and *Goveclop* (government close to the people).

When the Museum of Modern Art in Manhattan had a big exhibit of pop art, back in the days when pop was the latest art craze, did any newspaper think to headline a story "MOMA Shows Pop"? Come to think of it, in earlier days when the museum exhibited dada art, MOMA certainly showed dada.

Does the Engineering Information External Inquiries Officer of the BBC, when he answers the telephone, open with "EIEIO"?

In March 1983 the Eastman Kodak company suddenly realized that its newly formed U.S. Equipment Division had the acronym USED. On the assumption that nobody wants to buy used equipment, they sensibly renamed it the U.S. Apparatus Division.

The Boston Redevelopment Authority is obviously devoted to the uplift of Boston. And have you heard of IBTA, an organization opposed to topless swimsuits? The letters stand for the Itty Bitty Titty Association.

"What is the speediest reply to a boring remark?" writes Stephen Barr. The answer, he says, is OOMPH (Over One Mile Per Hour).

It is well known that NEWS is an acronym of North, East, West, South. So is SNEW. What's *snew?* Not much. What's new with you? Not well known is the startling fact that ADAM uses the initial letters of the Greek words for north (Arktos), west (Dusis), east (Anatole), and south (Mesembria). And did you know that Adam and Eve were Irish? When they first met, each lifted up the other's fig leaf. "O'Hair!" shouted Adam. Eve replied with "O'Toole!"

Riddles

When I was a boy I invented the following riddle: How did the man with big feet put on his pants? Answer: over his head. To my chagrin, I later discovered that the Reverend Edward Lee Hicks had recorded in his diary: "Heard this evening the last new joke of the author of *Alice in Wonderland:* He (Dodgson) knows a man whose feet are so large that he has to put on his trousers over his head."

The only other riddle I ever invented, which I believe no one beat me to, is this. Who was our tallest president? Answer: Dwight D. Eiffeltower.

There are hundreds of similar riddles that pun on well-known names. What weighs six tons and sings calypso? Harry Eliphante. What's green and dances? Fred Asparagus. Why is a martini without an olive or lemon twist called a Charles Dickens? No olive or twist.

His father was Japanese and his mother was Jewish. What did he do on December 7? He attacked Pearl Schwartz.

Who speaks softly and carries a big stick? The usual answer is a gay policeman or a pole vaulter, but I thought of a better one: Don Juan.

The bun, someone said long ago, is the lowest form of wheat.

Formula Jokes

In my *February 1981 Kickshaws,* I quoted five answers to the old riddle "What's black and white and red all over? " published in John Allen Paulos's *Mathematics and Humor,* and six more were given in Word Ways' May 1981 "Colloquy" column. Paulos's are from a much longer list given by M. E. Barrick in his paper "The Newspaper Riddle Joke," published on pages 253–57 of the 1974 volume of the *Journal of American Folklore.*

Has anyone compiled a similar list of answers to "Why did the chicken cross the road?" Here are six from Mary Ann Madden's book *Son of Giant Sea Tortoise:* because it was there, to get away from Colonel Sanders, because of an alternate-side parking rule, to avoid a street demonstration, because she did not want to get involved. My favorite answer is, to keep its pants up.

Matt Freedman and Paul Hoffman had a book published in 1980 titled *How Many Zen Buddhists Does it Take to Screw in a Light Bulb?* The book consists entirely of variants on this question. You may not know that the original riddle is very popular in Poland, where it is phrased: How many Americans does it take to screw in a light bulb? The answer is, one.

I would like to see a similar book on variations of "Waiter, there's a fly in my soup" and "Who was that lady I saw you with last night?" I collect versions of both jokes. Some of the lady-wife variants are based on wordplay. Who was that lady I saw you out with last night? I wasn't out, I was just dozing. Who was that lady I saw you outwit last night? Magician: who was that lady I sawed with you last night? Who was that ladle I saw you with last night? That was no ladle, that was my knife. Who was that hobo (or strumpet) I saw you with last night? That was no oboe (or trumpet), that was my fife.

Help may be on the way. The editor of *Word Ways* informs me that Paul Dickson, the author of *Words* (reviewed in the November 1982 issue of *Word Ways*), is now planning a book on formula jokes of all types. If it sees the light of day, I will be the first to buy a copy.

The Integers Revisited

In my *February 1981 Kickshaws,* I presented a puzzle in which ten
students in a class had the integers concealed in their names: dON
Edwards, roberT WOrden, etc. Cynthia Knight of Chicago, Illinois,
utilized the same device in an imaginary bit of cocktail-party conversation:

> No, never!
> That wouldn't do?
> It might be worth reexamining.
> Or else it's the end of our friendship.
> If I've understood you right, you've read my mind.
> Yes, I X-rayed it.
> That's even worse.
> I'll weight your remark.
> You see confusion in everything.
> That ends it!
> I feel even worse now.

Postscript

Charles Suhor (*Word Ways,* February 1984) added to my list of unintended puns the following lines from Robert Frost's famous poem
"Mending Wall":

> Before I built a wall I'd ask to know
> What I was walling in or walling out,
> And to whom I was like to give offence [a fence].

David Shulman, in the same issue of *Word Ways,* reported that
he had found the joke about the man with big feet in *The New-
Yorker,* Vol. 6, 1838, on page 88/3: "Sam Slick says he knew a man
down East whose feet were so big that he had to pull his pantaloons
over his head."

Paul Dickson's *Jokes* was published by Delacorte Press in
1984. The first chapter has forty-three pure examples of "fly in my
soup" jokes and sixty-nine variations. Chapter 20 covers black,
white, and red all over; who was that lady?; and the chicken that
crossed the road. It is a marvelously funny book. So is his later book
Names (Delacorte, 1986), described on the jacket as "A Collector's
Compendium of Rare and Unusual, Bold and Beautiful, Odd and
Whimsical Names."

Seven Puzzle Poems

Puzzlesmiths like to make things difficult for themselves. The ordinary constraints of language—grammar and meaning—are not tough enough, so they invent special rules: a grid that is filled with intersecting words, a sentence that reads the same backward and forward, a paragraph that is written using only the words in the Pledge of Allegiance.

A poet has similar instincts. Though his primary concern is refining sound and sense, he makes his job harder by self-imposing schemes of rhyme, meter, or alliteration.

That a poet should turn puzzler, or vice versa, is a natural transformation—and one with a long history. Almost 2,500 years ago, the Greek poet Pindar wrote an ode without using the letter sigma; another Greek poet, Tryphiodorus, composed a twenty-four-volume epic about Ulysses, each book omitting one letter of the Greek alphabet.

Hundreds of years later, in fifteenth-century Persia, the renowned poet Jami was approached by a lesser poet who wanted to read the great man a rhyme he had written.

"This work is quite unusual," the lesser poet proudly stated when he was done reading. "The letter *aliff* is not to be found in any of the words!"

"You can do better yet by removing *all* the letters," was Jami's curt rejoinder.

The seven puzzle poems here each represent a particular type of wordplay. Can you determine what is remarkable about the structure of each poem?

This article is reprinted from *Games* Magazine (810 Seventh Avenue, New York, NY 10019). © 1984 PSC Games Limited Partnership.

1. Square Poem

I often wondered when I cursed,
Often feared where I would be—
Wondered where she'd yield her love,
When I yield, so will she.
I would her will be pitied!
Cursed be love! She pitied me . . .

—Lewis Carroll

2. Capacity

Capacity 26 Passengers
—sign in a bus

Affable, bibulous,
corpulent, dull,
eager-to-find-a-seat,
formidable,
garrulous, humorous,
icy, jejune,
knockabout, laden-
with-luggage (maroon),
mild-mannered, narrow-necked,
oval-eyed, pert,
querulous, rakish,
seductive, tart, vert-
iginous, willowy,
xanthic (or yellow),
young, zebuesque are my
passengers fellow.

—John Updike

3. Curious Acrostic

Perhaps the solvers are inclined to hiss,
Curling their nose up at a con like this.
Like some much abler posers I would try
A rare, uncommon puzzle to supply.
A curious acrostic here you see
Rough hewn and inartistic tho' it be;
Still it is well to have it understood,
I could not make it plainer, if I would.

—Anonymous

4. I Will Arise

I
will
arise
and
go
now,
and
go—any damned place
    just to get away from
        THAT
        chair
        covered
        with
        CAT
        hair

—William Jay Smith

5. Winter Reigns

Shimmering, gleaming, glistening glow—
Winter reigns, splendiferous snow!
Won't this sight, this stainless scene,
Endlessly yield days supreme?

Eyeing ground, deep piled, delights
Skiers scaling garish heights.
Still like eagles soaring, glide
Eager racers; show-offs slide.

Ecstatic children, noses scarved—
Dancing gnomes, seem magic carved—
Doing graceful leaps. Snowballs,
Swishing globules, sail low walls.

Surely year-end's special lure
Eases sorrow we endure,
Every year renews shared dream,
Memories sweet, that timeless stream.

—Mary Hazard

6. Night's Pilgrim

Idling, I sit in this mild twilight dim,
Whilst birds, in wild, swift vigils, circling skim.
Light winds in sighing sink, till, rising bright,
Night's Virgin Pilgrim swims in vivid light!

—Anonymous

7. Spa

Laughing boys, legs bare, with girls bathing—
　Girls kind of fond are these,
Chaffing and cheering boys, limbs writhing..
　　Swirls water, whips spume, splash seas,
　　　Breaking, into shrieking
　　　　Girls . . .
　　　　Noise and boys,
　　　　Boys and noise . . .
　　　　Girls,
　　　Shrieking, into breaking
　　Seas splash, spume whips, water swirls..
Writhing limbs, boys cheering and chaffing—
　These are fond of kind girls,
Bathing girls with bare legs, boys laughing.

—J. A. Lindon

Answers

1. This "square" poem reads the same forward as it does when the first word of each line is read in order, followed by the second word of each line, etc.

2. The initials of the first twenty-five words (not counting the words in parentheses) are the letters of the alphabet in order—minus the letter *U*. According to the sign quoted at the beginning of the poem, the bus holds twenty-six people. Updike is describing only his *fellow* passengers, so the poet himself is the missing *U*.

3. The first two letters of each line, read in sequence, spell "peculiar acrostic."

4. The poem is in the shape of its subject, a chair.

5. The last letter of each word, including the words in the title, is the first letter of the following word.

6. The only vowel in this poem is *I*.

7. This is a word palindrome; it reads the same forward as it does when the words are read in reverse order.

Slicing Pi into Millions

> If we take the world of geometrical relations, the thousandth decimal of pi sleeps there, though no one may ever try to compute it.
> —William James, *The Meaning of Truth*

In 1909, when William James doubted that pi[1] would ever be computed to a thousand decimals, the record for such a calculation was held by an obscure nineteenth-century British mathematician named William Shanks. He had worked out pi to 707 decimals, and for more than seven decades no one bothered to check his figures.

Poor Shanks. He had spent twenty years doing his calculations by hand—with probably nothing more than a crude mechanical calculator to help him—only to fumble after 527 correct decimals. The 528th is 4, but Shanks called it 5, and from there on his digits are wrong. The error went undetected until 1945, when another Englishman, D. F. Ferguson, discovered it. Four years later the value of pi was accurately extended to 1,120 decimals by two Americans, John W. Wrench, Jr., and Levi B. Smith, in what turned out to be the last effort to compute pi on a preelectronic desk calculator.

As it happens, the thousandth decimal of pi is 9. The number in itself is not important, but its discovery raises a question so profound that philosophers and mathematicians strongly disagree on the answer. The question: Was the first sentence of this paragraph true *before* the 1949 calculation? To those of the realist school, the sentence expresses a timeless truth whether anyone knows it or not. In their view, what happened in 1949 was not that it suddenly became true but that human beings discovered its truth. Not so, say philosophers and mathematicians of nonrealist persuasion, whose views are close to the pragmatism of William James. They prefer to

---

This article originally appeared in *Discover*, January 1985, and is reprinted here, with changes and a postscript, with permission. Copyright 1985 by Discover Publications, Inc.

1. In case you have forgotten your basic geometry, pi is the ratio of the circumference of a circle to its diameter. One well-known formula for calculating it is $\pi = 4/1 - 4/3 + 4/5 - 4/7 + 4/9 - \ldots$

think of mathematical objects as having no reality independent of the human mind. (We leave aside the possibility that extraterrestrials may have calculated pi to a thousand decimals before Smith and Wrench did.)

James defended a middle view. The uncalculated decimals of pi, he said, "sleep in a mysterious abstract realm where they have a pale sort of reality. Not until they are calculated do they become strongly real, and even then their reality is merely one of degree. The first calculation of the thousandth decimal could have been as mistaken as Shanks's 528th. Only when later calculations confirmed it did the 9 become, in the Jamesian sense, more fully awake. Today no one has reason to doubt that the thousandth decimal is 9. Since the search for further values of pi has been largely computerized, it has presumably been freed of arithmetical errors.

The first computer to tackle pi was ENIAC (for electronic numerical integrator and computer), which carried its value out to 2,037 places. This dinosaur took seventy hours to complete the job. Five years later NORC (naval ordnance research calculator) computed pi to 3,089 places, and in 1957 Wrench and Daniel Shanks (no relation to the erring William), using an IBM 7090, ran pi to 100,265 decimals in eight hours and forty-three minutes. In 1973 the French mathematician Jean Guilloud reached a million, a calculation that took twenty-three hours and eighteen minutes on an IBM 7600. The French atomic energy commission considered the results important enough to publish as a four-hundred-page book.

Is a million digits the record? Not by a long shot! In 1983 the University of Tokyo's Yoshiaki Tamura and Yasumasa Kanada, using the superfast HITAC M-280H computer, stretched pi to 2^{24}, or 16,777,216, places in less than thirty hours. In 1984 these results were verified on an even faster computer, to 10,013,395 decimals, the accepted record at the moment. Kanada and his associates are planning to go to 2^{25}, or 33,554,432, digits, and eventually to more than 100 million. Will the billionth decimal ever be aroused from its deep slumber? Possibly. In fact, someone may determine it without having to calculate all the preceding digits, although so far no one has any idea how this could be done.

The Tokyo results are all based on a remarkable algorithm, a systematic calculating procedure, devised a decade ago by Eugene Salamin at MIT. The algorithm is based on an infinite series of fractions that when extended converge with great rapidity on pi. The number of calculated digits doubles at each step, which explains why the Tokyo figures are powers of 2. At first, Salamin thought his series was original. Then he learned he had rediscovered a formula published in 1818 by the German mathematical genius Carl Friedrich Gauss. No one had considered using it for computing pi because it involved such time-consuming multiplication. Only with the advent of high-speed supercomputers and clever new procedures for multiplying has it become practical to put Salamin's (or Gauss's) algorithm to work calculating pi.

Still, even with the electronic help, why should anyone bother to carry pi to such fantastic lengths? There are four reasons:

1. Pi is *there*—wherever "there" is!

2. Such calculations have useful spinoffs. Much is learned about calculating and checking large numbers on computers.

3. The calculation of pi to tens of thousands of places provides useful exercises for testing new computers and for training programmers.

4. The more digits of pi that are known, the more mathematicians hope to answer a major unsolved problem of number theory: Is pi's sequence of digits totally patternless, or does it exhibit a persistent, if subtle, deviation from randomness?

To explain a random sequence of digits, mathematicians offer an analogy: Imagine yourself at a gaming table betting on the next digit to appear while a sequence of digits is being generated, perhaps by a roulette wheel. If there is no possible way of predicting the next digit with a probability of better than one in ten, then the sequence is random. In this sense, pi is certainly not random. You can always make your own calculation and predict the next digit with certainty.

In another sense, however, pi can be called random. As far as anyone knows, it shows no trace of a pattern, no kind of order, in the overall arrangement of its digits. This curious property is shared with the square root of 2 and an infinite number of other irrational numbers (numbers that cannot be expressed as fractions with integers above and below the line). Every digit has the same probability (one in ten) of appearing at any one spot, and the same conformity to randomness applies to so-called doublets, triplets, or any specified pattern of digits, adjacent or separated, in pi's endless stream of digits. Herein lies the rub: no one has proved that pi is patternless, nor has anyone proved it is not. No one has even proved that each digit must appear in pi an infinite number of times.

Yet, even if we assume that pi is patternless, it does not follow that pi does not contain an endless variety of remarkable finite subpatterns that are the result of pure chance. For example, starting with pi's 710,100th decimal is the stutter 3333333. Another run of seven 3's starts with the 3,204,765th decimal. There are runs of the same length, among the first ten million decimals of pi, of every digit except 2 and 4. Digit 9 leads with four such runs; 3, 5, 7, and 8 each have two seven-runs; and 0, 1, and 6 have one run apiece. There are eighty-seven runs of just six repetitions of the same digit, of which 999999 is the most surprising because it comes so soon, relatively speaking: it starts with the 762d decimal.

The ascending sequence 23456789 begins with decimal 995,998, and the descending sequence 876543210 starts at decimal 2,747,956. Among the first ten million decimals of pi, the sequence 314159—the first six digits of pi (which were known as long ago as the fifth century A.D. by Chinese mathematicians)—appears no fewer than six times. The first six digits of e, a celebrated number in mathematics that can be defined as the basis of natural logarithms, occurs eight times, not counting one appearance (at decimal 1,526,800) of 2718281, the first seven digits of e. Even more unexpected is the appearance (starting with the 52,638th decimal) of 14142135, the first eight digits of the square root of 2.

There is a lesson to be learned from these strange coincidences. Consider the 876543210 pattern. The probability of finding it among pi's first three million digits is low—about six chances in 100. But the probability that *some* improbable patterns will turn up is extremely high.

We can look for still other oddities in pi. If the first n digits of pi form a prime number (a number divisible only by itself and 1), let us call it a pifor (pi forward) prime. Only four such numbers are known: 3, 31, 314159, and 3 141549, and 3 14159 26535 89793 23846 26433 83279 50288 41. The fourth was proved to be a prime in 1979 by Robert Baillie and Marvin Wunderlich of the University of Illinois. Is there a fifth? Probably, but it could be a long time before anyone knows.

What about piback primes, or the first *n* digits of pi running backward? We would expect them to be more numerous than pifors because all pibacks end in 3 (the first digit of pi), one of four numbers that a prime must end with; the others are 1, 7, and 9. By contrast, pifor numbers can end in any digit, which means only 40 per cent of the numbers have a chance to be prime.

Six pibacks can be easily identified: 3, 13, 51413, 951413, 2951413, and 53562951413. Now, thanks to a calculation by Joseph Madachy, editor of the *Journal of Recreational Mathematics,* we know that 979853562951413 is prime as well. Baillie reports that there are no other piback primes through the 432rd decimal of pi. Sharp-eyed readers may note that the first three pifors are in fact reversals of three pibacks. Is there a larger prime that belongs in this way to both sets? Perhaps.

True numbers nuts (I use the term affectionately) will probably ask still another question: Do the first n digits of pi ever produce a pifor square number that is, a number that is the square of another number, as 4 is of 2, or 9 of 3)? Pibacks are ruled out because no perfect square can end in 3. University of Illinois mathematician Wolfgang Haken doubts that there are such squares. His reason: the further you go in the decimal expansion of pi, the less likely it becomes that a pifor square will be encountered. Haken estimates that the probability is as low as one in a billion. His conjecture may be true, but it may also be undecidable, because such a square—or any proof that it cannot "exist"—may never be found.

The quotes around *exist* indicate that we are back again in the metaphysical quicksands. Is it legitimate to say that 0123456789 *now* either sleeps in pi—or that it does not? A realist would reply: "Of course!" But nonrealists would disagree. If such a sequence is ever found, naturally that will settle the matter. But since it has not been, some mathematicians refuse to declare that it is or is not there. However, suppose we alter the assertion to "0123456789 either is or is not sleeping among the yet uncalculated first billion decimals of pi." All mathematicians would agree that this statement is indeed true. Why? Because now it can be settled conclusively in a finite number of calculations. Even if pi has to be sliced into a billion pieces.

Postscript

In the fall of 1985 R. William Gosper, Jr., of Symbolics, Inc., a small firm in Palo Alto, California, calculated pi to seventeen million digits by using continued-fraction expansions of his own devising and nothing more than a small desk computer. In January of the following year this record was superseded by pi to 29,360,128 digits. Using a program written by David Bailey and based on an algorithm discovered by Jonathan Borwein and Peter Borwein, a Cray-2 supercomputer finished the job in twenty-eight hours. The record was short lived. In 1986 the University of Tokyo group extended the expansion to 134,217,700 digits, and in 1987 they carried pi to 201,326,000 digits.

When I said no "patterns" had been found in pi I was cutting corners with an extremely fuzzy word. More precisely, pi has so far passed all tests for what is called a "normal" number. This means that when pi is expressed in any base notation, no one has yet found in its expansion any significant deviations from random expectation for any digit or finite string of digits. With respect to decimal notation this means that each digit appears with a frequency of 1/10, each pair of digits with a frequency of 1/100, each triplet with a frequency of 1/1,000, and so on. A normal number must be irrational (because all rational decimal expansions have a repeating string of digits), and almost all real numbers are normal. Put another way, the numbers not normal have what mathematicians call a "measure zero."

A normal number may be strongly patterned. A famous example is the decimal fraction obtained by putting down all the counting numbers in counting order (.1234567891011121314151617. . .). It has been proved normal, but it cannot be called unpatterned. Nobody knows whether pi, e, or any irrational root of an integer is normal.

Pi is not, of course, patternless in a wide sense, because it is the limit sum of simple infinite sequences of fractions. The word "pattern" seems to me to have a continuum of meanings, with respect to sequences of digits, that stretches from the obvious pattern of the decimal expansion of 1/3 to the deeply concealed "pattern" of pi that derives from any algorithm used for calculating it.

Here is a delightful poem by Andrew Lang that I recently came across:

Ballade of a Girl of Erudition

She has just put her gown on at Girton.
She is learned in Latin and Greek;
But lawn tennis she plays with a skirt on
That the prudish observe with a shriek.
In her accents perhaps she is weak
(Ladies *are,* one observes with a sigh),
But in her algebra—there she's unique,
But her forte's to evaluate π.

She can talk about putting a "spirt on"
(I admit an unmaidenly freak),
And she dearly delighteth to flirt on
A punt in some shadowy creek;
Should her bark by mischance spring a leak,
She can swim as a swallow can fly;

She can fence, she can put with a cleek,
But her forte's to evaluate π.

She has lectured on Scopas and Myrton,
Coins, vases, mosaic, the antique,
Old tiles with the secular dirt on,
Old marbles with noses to seek,
And her Cobet she quotes by the week,
And she's written on *Ken* and on *Kai,*
And her service is swift and oblique,
But her forte's to evaluate π.

Envoy.

Success like a rose is her cheek,
And her eyes are as blue as the sky;
And I'd speak had I courage to speak,
But her forte's to evaluate π.

Along with my article I sent *Discover* a set of curiosities involving pi that I selected from a large collection. *Discover* published a few of them. Here is a complete list:

Numbers That Get Curiouser and Curiouser

In the fifth century A.D. the great Chinese astronomer Tsu Ch'ung found a remarkable fraction: 355/113. It gives pi correct to six decimals. Western mathematicians did not discover this value until a thousand years later. If you write the fraction backward, altering only one digit, 553/312, you get 1.7724358+, which is the square root of pi correct to four decimals.

The square root of 10 is pi correct to one decimal. The square root of 2 plus the square root of 3 is correct to two decimals. The cube root of 31 is correct to three decimals. The square root of 9.87 (note the reverse counting order) is correct to a rounded four decimals. The square root of 146 times 13/50 is correct to a rounded six decimals.

Try this on your calculator. Divide 2,143 (the first four counting numbers) by 22, then hit the square-root button twice. You will get pi correct to eight decimals. This astonishing formula, $22\pi^4 = 2,143$, was discovered in 1914 by the famous Indian mathematician Srinivasa Ramanujan.

It is easy to prove that pi are square (get it? πr^2, the formula for calculating the area of a circle). Remember that pi is the sixteenth letter in the Greek alphabet and that 16 is the square of 4.

In the English alphabet, let $A = 1$, $B = 2$, and so on. P again has a value of 16, and I has a value of 9, the square of 3. The sum of 16 and 9 is the square 25, and the product is the square 144. Divide 9 by 16 and you get a decimal fraction with the repeating period 5625, the square of 75.

All pies are square because the letters of *PIES* add up to 49. Pie à la mode = 81, raisin pie = 100, coconut pie = 121, and Eskimo pies = 121. Are there other square pies?

The first 144 decimals of pi add up to 666, the New Testament's notorious number of the Beast, or anti-Christ (Revelation 13:18). Note that $144 = (6 + 6) \times (6 + 6)$. The three decimals of pi that begin with the 666th are $343 = 7 \times 7 \times 7$.

Shown in figure 9.1 are the capital letters of the alphabet. Cross out all of those with left-right symmetry (letters that look the same in a mirror). The remaining letters form groups whose number of letters, taken clockwise, gives 31416.

The best integer-fraction approximation of pi that uses each of the ten digits just once is 67389/21450. It is correct to a rounded four decimals.

A circle has 360 degrees. This number is the triplet in pi that ends on the 360th decimal.

Pie is an obsolete spelling of pi. Hold the word, as printed in figure 9.2, up to a mirror. You will see the first three digits of pi.

From a finite set of the first n counting numbers, two are si-

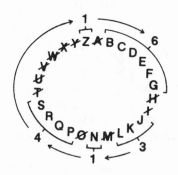

Figure 9.1

Figure 9.2

multaneously selected at random. What is the probability that they are relatively prime (have no common divisor)? As n increases, the probability rapidly converges on the square root of pi.

Pi has no 0 until decimal 32, and pi squared has no 2 until decimal 47.

Albert Einstein was born on 3/14.

"Which transcendental number do you like best, pi or e?" "I prefer pi," she replied palindromically.

Underwood Dudley, a DePauw University mathematician, mentions this curiosity in a fine article on "What to Do When the Trisector Comes," in *The Mathematical Intelligencer,* vol. 5 (1983), on pages 20–25:

Arrange the six permutations of 1, 2, 3 in ascending order and take their first differences:

$$
\begin{array}{cccccc}
123 & 132 & 213 & 231 & 312 & 321 \\
& 9 & 81 & 18 & 81 & 9
\end{array}
$$

The top row adds to 1,332, the bottom row to 198. Arrange the first nine digits of pi in triplets and add 198 to each:

$$
\begin{array}{ccc}
314 & 159 & 265 \\
198 & 198 & 198 \\
\hline
512 & 357 & 463
\end{array}
$$

The three totals add to 1,332.

10   The Traveling Salesman

A traveling salesman wants to visit ten towns on a cross-country swing. It's fairly easy for him to arrange his itinerary so that he clocks the least amount of mileage-and reduces his travel time and expenses. But if the number of towns increases significantly (say, to five hundred or more), so does the complexity of his travels. Even a computer might have to wrestle for years—or perhaps centuries—to find the shortest routing among the towns.

This classic problem has not only been the death of salesmen (figuratively speaking) but also of mathematicians. Despite the best efforts of some of the world's most agile minds, it has defied easy solution. In fact, it is only one of thousands of similar unsolved mathematical brain teasers, many of which have urgent applications in the efficient operation of science, technology, and industry. All belong to a new, rapidly expanding field of computer science called complexity theory.

Complexity, in this case, is a measure of how long it takes a computer to solve a particular problem. As the number of variables increases, so does complexity. Because time is costly on the big mainframes used to crack such problems, computer scientists strive to develop the fastest, most efficient possible algorithm— the step-by-step instructions used by the machine to tackle the problem.

In 1984 Narendra Karmarkar, a young Indian-born mathematician at AT&T's Bell Laboratories, developed a startlingly improved method for doing what computer specialists call linear programming—a technique for handling thousands of equations simulta-

This article originally appeared in *Discover,* April 1985, and is reprinted here, with changes and a postscript, with permission. © 1985 by Discover Publications, Inc.

neously. Appropriately enough, Karmarkar's algorithm has an immediate usefulness in the telephone business: it will help route millions of calls between millions of locations, possibly saving millions of dollars a year. But it is also expected to provide important benefits to other businesses as well, including those that must dispatch people or equipment around the country.

As it happens, the traveling-salesman problem belongs to a particularly stubborn class of mathematical conundrums known as NP complete, first identified by computer scientists in the 1970s. (If you really want to know, NP stands for nondeterministic polynomial.) So far, these problems have resisted all efforts to find fast, efficient algorithms to solve them—not even Karmarkar's does more than reduce the computing time for getting an approximation of a solution. Yet, they are so closely related that if a good algorithm is found for just one, it will apply immediately to all the others as well—and solve all outstanding NP-complete problems in one swoop.

To understand what computer scientists mean by "good" or "fast" algorithms, we must be able to distinguish between what they call polynomial time and exponential time. A diversion into the recent history of mathematical games may be helpful. One of the most captivating geometrical recreations involves sets of what Solomon Golomb, at the University of Southern California, calls polyominoes. He introduced the concept in a paper he wrote in 1954 as a twenty-two-year-old student at Harvard. A polyomino is a figure formed by joining unit squares at their edges. Figure 10.1 shows the single monomino, the single domino, the two trominoes, the five tetrominoes, and the twelve pentominoes. So many challenging problems arise from these shapes that Golomb wrote a book called *Polyominoes.* One reader, Arthur Clarke *(2001),* used pentominoes in his novel *Imperial Earth* as a symbol of the combinatorial possibilities of life. Clarke's latest book, a collection of essays called *Ascent to Orbit,* includes a chapter entitled "Help—I Am a Pentomino Addict!"

Among the endless unanswered polyomino questions, the most fundamental is this: Is there a formula that gives the number of different polyominoes that can be made with n squares? There are computer algorithms for calculating the number, but they are not efficient, because they rely on so-called recursive methods. To find all the polyominoes of order n, the computer must first know all the polyominoes of order $n - 1$, then test all possible ways of adding a unit cell to each, and then eliminate the duplicates. For

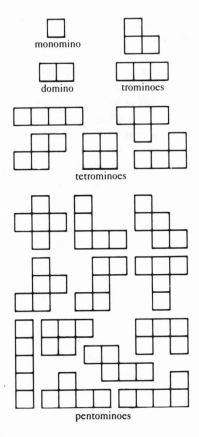

Figure 10.1: Polyominoes

n = 6 through 18, the number of polyominoes (mirror reflections are not counted, but shapes with interior holes are included) are 35, 108, 369, 1285, 4655, 17073, 63600, 238591, 901971, 3426576, 13079255, 50107911, and 192622052. Formulas are known that give upper and lower bounds, but none is known that pins down the exact number. The problem is also unsolved for shapes made when identical equilateral triangles or regular hexagons are joined. It is not yet known whether these problems are NP complete.

If pentominoes are new to you, you might enjoy making a set of the twelve pieces and seeing how long it takes you to fit them together like a jigsaw puzzle to make the T-pattern shown in figure 10.2. Asymmetrical pieces may be placed with either side up. There is only one solution. (See Answers section at end of chapter.)

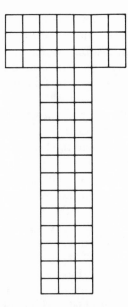

Figure 10.2: T-pattern of Pentominoes

Suppose, for instance, you have an algorithm for finding all polyominoes of the order n. The increase in time it takes as the computer goes up the ladder of n's is called polynomial if it is expressed by an algebraic formula in which n is not an exponent. Such algorithms are called "good" or "efficient" because the computing time, as the task gets more complex, increases relatively slowly. On the other hand, if the time growth is given by a formula in which n is an exponent, the time "explodes" by growing at an exponential rate. Such algorithms are "bad" or "inefficient." They quickly demand "unreasonable" computing time.

The two growth rates are illustrated in this fable: A boy asked his father on Christmas to give him an allowance of a penny on the first day of the new year, four pennies on the second day, nine on the third, and so on for the rest of the year. Each day's amount would be n^2, where n is the number of the day. The father calculated that on the 365th day he would have to give his son 365^2, or $1,332.25, so he refused. As a counterproposal, the wily lad suggested a penny on the first day, two on the second, four on the next, eight on the next, and so on in a series that doubled each day, with the allowance ending on the last day of January. After January, said the boy, he would never ask for a penny again. The daily amount is now given by 2^{n-1}. This seemed much more reasonable to the father. After all, eight pennies on the fourth day is half the sixteen required by the former scheme, and a month is much shorter than a year. The hapless father agreed. He had no inkling that on the 31st day he would have to give his son 2^{30}, or $10,737,418.24.

The distinction between polynomial and exponential computing time is admittedly fuzzy, because the time varies with both how an algorithm is programmed and the kind of computer used, but the main idea is this: if an algorithm is polynomial, you can go a long way up the n ladder with only a moderate growth of computer time; but if the algorithm is exponential, you can quickly reach a value of n that would require thousands of centuries of computer time to solve the problem exactly. In computer science a problem is not considered "well solved" unless a polynomial-time algorithm is known.

Now let us return to the most famous NP-complete problem, that of the traveling salesman, and look at it mathematically. A salesman wants to make one visit to each of n towns, starting and ending at the same town. What is the shortest path he can take? When n is small, a computer can examine all possible routes and pick the shortest; but as n increases, the number of possible routes grows exponentially and there are no known algorithms that can find the shortest path in polynomial time. Lots of ingenious algorithms do a good job of getting very close to the minimal path in a reasonable time, but finding the precise path remains elusive. Fig-

Figure 10.3: By traveling this route, a salesman could visit all forty-eight state capitals and clock the lowest possible mileage.

ure 10.3 shows a solution to the traveling-salesman problem for the forty-eight capitals of the contiguous states (see Postscript for comments).

Closely akin to the traveling-salesman problem is the problem of deciding whether an arbitrary graph has what is called a Hamiltonian path (after the nineteenth-century British mathematician Sir William Hamilton). A graph is simply a set of points connected by lines. If you can start at one point and traverse the lines so as to visit each point just once, you will have followed a Hamiltonian path. If the path returns to its starting point, it is called a Hamiltonian circuit. Finding such paths and circuits is an NP-complete task even when a graph is "directed" (one on which you can traverse each line only in the direction indicated by an arrow). Henry Dudeney, England's greatest puzzlemaker, has many Hamiltonian-path puzzles in his books. Figure 10.4 shows one of the easiest.

Almost as famous as the traveling-salesman problem is another NP-complete task, known as the subset-sum problem. It is the simplest version of what are often called knapsack problems—an analogy to packing a knapsack to meet certain provisos. Knapsacks

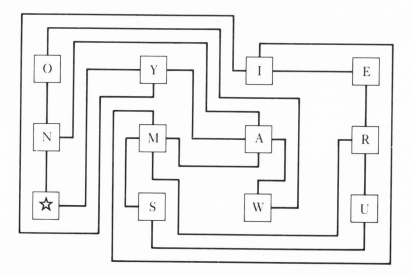

Figure 10.4: Two cyclists start at the square marked with a star. They want to travel the roads so as to visit each of the eleven towns once and only once, ending the tour at E. "I'm certain it can be done," says one cyclist. "No way, I'm sure," says the other. Who is right? (See Answers section at the end of the chapter for the solution.)

have been in the news recently because several new cryptographic systems—some of which are used by the military, others of which are used by banks for the electronic transfer of funds—are based on them. The term *knapsack* is most often applied to a problem involving a set of objects each of which has two values—say, size and weight. Think of the size as shown by a blue number on the object, the weight by a red number. You seek a subset of objects such that the sum of their blue numbers will not exceed a specified value (the knapsack's capacity), and at the same time you want to maximize the sum of their red numbers. This in turn may be thought of as a special case of the more general "bin packing" problem, in which the knapsack is replaced by a finite number of bins.

Like Hamiltonian paths, versions of subset-sum abound in the literature of recreational mathematics. They frequently take the form of a target with concentric rings, each ring having an integer (whole-number) value. The object is to shoot arrows or bullets at the target to obtain a specified total score. Figure 10.5 shows how Sam Loyd, America's counterpart of Dudeney, dressed up a subset-sum problem in his 1914 *Cyclopedia of Puzzles*.

Loyd's puzzle is easily solved by trial and error, but such problems are far from trivial when there are lots of big numbers. Moreover, packing problems are handled by computer algorithms that have a surprising variety of applications to loading cargo, warehouse storing, budget planning, and many other industrial tasks. Apart from such utility, the solving of difficult problems is the driving force behind the progress of pure mathematics, and without it we would have neither science nor industry.

Indeed, some of the unsolved math problems are so significant and deep that a solution would be a major world event (well, at least in mathematics circles). Others remain unsolved or even forgotten mainly because mathematicians consider them too dull or trivial to justify working on them. Still others lie within the grasp of even those of us who are not mathematics prodigies. But let us take a look at the greatest unsolved problem of all. Consider this simple equation: $a^n + b^n = c^n$. If $n = 2$, it has an infinity of solutions in integers. The simplest is $3^2 + 4^2 = 5^2$. Such solutions are called Pythagorean triples because they measure the sides of right triangles. Are there solutions in integers if n is greater than 2? The famous seventeenth-century French mathematician Pierre Fermat scribbled in the margin of a book that he had discovered a "truly marvelous" proof that the answer is no, adding that the margin was too small to contain it. Fermat never disclosed his proof. Most mathematicians think he later discovered it was faulty.

Figure 10.5: A subset-sum problem from Sam Loyd's vintage *Cyclopedia of Puzzles:* Can you knock down a set of dummies whose numerical values will add up to 50?

To this day no one has been able to prove Fermat's last theorem, as it came to be known, or to find a solution that falsifies it. Do not waste time looking for a counterexample! Its exponents would have to exceed 125,000, and the values of *a, b,* and *c* would have to be millions of digits long.

Fermat's conjecture is probably true, but it could be what mathematicians call an "undecidable" true theorem. If so, the prospects for proving it are bleak. For centuries mathematicians will struggle vainly to construct a proof, and of course they will never find a counterexample. If the theorem is false, of course, it cannot be undecidable, because a single counterexample would decide it. In 1983 a young West-German mathematician, Gerd Faltings, made some progress by proving that if the theorem is false, then the equation has only a finite number of basically different solutions for each exponent.

Among hundreds of unsolved problems in number theory, one of the most curious is the palindrome conjecture. Write down any number of more than one digit. Reverse it and add it to the original number. Then reverse the sum, add again, and keep this up until you form a palindrome—a number that is the same in both directions. For example, suppose you start with 1985 and proceed as follows: 1985 + 5891 = 7876 + 6787 = 14663 + 36641 = 51304 + 40315 = 91619.

The palindrome is obtained in just four steps. For decades it was assumed that every starting number produced a palindrome. Charles Trigg, a retired mathematician now living in San Diego, was not so sure. He began testing numbers in consecutive order and found that all of them reached a palindrome in six or fewer steps until he came to 89. That one required twenty-four steps. Of course its reversal, 98, also took twenty-four steps. From there on, palindromes appeared in no more than twenty-four steps until Trigg reached 196. To his amazement, after 100 steps there was still no palindrome. He found 149 integers under 1,000 that seemed unable to generate a palindrome. By 1967 he was convinced that the palindrome theorem is false.

In 1975 Harry Saal, at IBM's Israel Scientific Center, used a computer to test 196 to 237,310 steps, reaching a final sum with 98,305 digits. Still no palindrome. No one had yet proved the existence of a number that would never reach a palindrome, although it is curious that such numbers have been shown to exist in any system of notation based on a power of 2. For instance, if you start with the binary number 10110 (or 22 in decimal notation), the sums fall into a four-step cycle that keeps extending a basic pattern

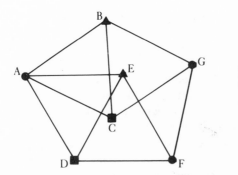

B & E = ▲
C & D = ■
A, G & F = ●

Figure 10.6

that is not reversible. The question remains unanswered for all other notation systems.

The notorious four-color map theorem—which says that every map can be colored with only four colors so that no two regions of the same color share a border—was finally proved in 1976 (by using a computer's brute-force number-crunching power!). But scores of other map-coloring problems resist all efforts to solve them, even with the help of powerful machines. Imagine an infinite plane divided into regions of different colors—and in such a way that no two points a unit distance apart are of the same color. What is the smallest number of colors needed for such a map?

Using the graph shown in figure 10.6 we can show that three are not enough. Each line of this graph has a length of 1. Let us assume that the graph is placed anywhere on a three-color map that solves the problem. If point A is on red, then B and C must be on the other two colors, and G must be on red. Similarly, D and E must be on the other two colors, and F must be red. But G and F are a unit distance apart, both red; therefore the original assumption is contradicted. No three-color map can have the desired property.

If the plane is tesselated in the manner shown at the bottom of figure 10.7 it is easy to see that seven colors (here represented by letters) will solve the problem. Each hexagon is slightly less than one unit from corner to corner. Can the problem be solved with six, five, or four colors? Nobody knows.

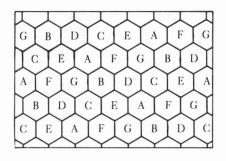

Figure 10.7

Answers

1.

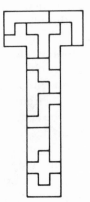

2. Both cyclists were right. The second cyclist was giving the answer. The problem is solved by taking the cities in the order that spells NO WAY I'M SURE.

3. 6,19,25.

Postscript

When this article appeared in *Discover,* the map showing the salesman's shortest route through the forty-eight mainland state capitals was incorrect, and I was swamped with letters from readers who found ways to improve it. Although the map had been published several times as a solution to the capitals problem, it turned out that it was based on a 1948 map on which many large cities had been substituted for capitals because accurate distances between certain capitals were not then available.

The problem must be carefully defined. Cities are fuzzy areas, not points. The shortest route by car is hopelessly vague because it depends both on what roads are open and on whether one intends to minimize distance or travel time. If the salesman goes by plane, airports may be far from the centers of cities.

At *Discover*'s request, the problem was given to Shen Lin, who directs the Network Configuration Department at Bell Labs, in Murray Hill, New Jersey. For the forty-eight points Lin chose the central AT&T office in each capital, then calculated the 1,128 distances (as the crow flies) between each pair of cities. (Mathematicians call this a "distance matrix.") The number of possible routes that allow the traveler to visit each city just once and return to the starting city, is 1/2 (48—1)! where "!" is the familiar factorial sign. It is a number of sixty digits. The difficult task—that is, difficult without computer—is to find the shortest of these routes.

Lin has a very fast "near optimal algorithm" of his own devising. Using a high-speed computer, Lin had his answer in less than a second. His map published in *Discover* (July 1985), is the map reproduced here. The length of the path is 10,628 miles.

Is it truly the shortest? Lin was so sure it is that he personally offered $100 to anyone who could find a shorter path based on his distance matrix for the forty-eight capitals. Since no one has yet claimed this prize, we can be confident that the problem has now— and for the first time—been laid to rest.

The Abacus

In China it is called a *suan pan,* and it has been a basic tool for centuries. Teachers there cannot be certified until they master its intricacies, and it is the universal tool of storekeepers. Students are drilled in its use in third and fourth grade, and they take advanced training if they go on to accounting, technical, or vocational schools. It is even the subject of a twenty-minute documentary film entitled *Home Town of the Abacus,* which is distributed throughout the country to promote its use.

In the Soviet Union, where it first appeared in the sixteenth century as an aid for computing taxes, it is called a *shchyoty.* Although no special effort is made to teach its use in schools, it is still essential in business for a very simple reason—the electronic calculator is all but unknown.

In Japan, where it is known as a *soroban,* its future is less certain. Fifteen years ago every storekeeper in Tokyo added up purchases and figured out change on an abacus. Today that storekeeper is much more likely to do his arithmetic on an electronic calculator. In 1970 only 1.4 million calculators were sold, but in 1983 their sales surpassed 68 million, while only two million abacuses were sold. In banks and accounting offices the rapid clicking of beads has given way to silent button pushing. Ten years from now an abacus may be as hard to find in Tokyo as a slide rule on the MIT campus.

But throughout the Far East, the abacus has legions of passionate partisans, simply because no other calculating instrument is more elegant, more accurate, or more fun to operate. In the hands of an expert it can perform addition and subtraction faster than an electronic calculator. In the hands of a master it will compute long division and multiplication problems with more digits than a hand calculator can accommodate. An abacus almost never wears out or needs repairing, and it is not dependent on electricity. In its own way it is as simple and beautiful as a burning candle. In offices, one sometimes sees an abacus behind glass on the wall by a giant computer with a sign that says, "In case of power failure, use this."

The history of this marvelous device spreads across much of the ancient world. One ancestor of the abacus was the counting board of Greece and Rome. It consisted of a slab of wood or marble, sometimes a piece of cloth or parchment, on which parallel lines were etched or drawn. The ancients did their calculations by moving pebbles or other counters back and forth on the lines. The Greeks called the device an *abakion,* from their word *abax,* a flat board; the Roman word was *abacus.* (Latin for pebble is *calculus,* the source of our word calculate. Medieval counting boards were often checkered, which explains the origin of such words as check and exchequer.) In Western countries the abacus has survived in the colored beads on playpens, in devices for teaching arithmetic to young children and the blind, and in counting aids such as the rosary and overhead beads for recording billiard scores.

What we call an abacus is no more than a counting board with its counters sliding in grooves or attached to parallel wires or rods. The Chinese probably originated the concept of beads on rods, but there are references to grooved devices in early Roman literature, and several Roman abacuses, similar in structure to the Japanese soroban, have survived. Calculations can begin at any of several points along such instruments, but in every case the counters on the rod selected to serve as the starting point stand for units of one. Each succeeding rod to the left represents the next higher power of ten. Each of the four counters below the horizontal bar, in the region the Japanese call earth, counts as one unit of the rod's value (ten, one hundred, and so on), while the single counter above the bar, in the region called heaven, equals five times that value. Only beads pushed up or down against the bar are counted.

With blurring speed, adroit abacus users flick the counters to the bar with the thumb and index finger. Tabulations on a soroban are always performed in a left (or highest value) to right sequence. (The procedure is reversed on a Chinese *suan pan.*) To add two numbers, a user slides markers of the value of the first number against the bar, then, beginning with the rod farthest to the left, adds to it the digits of the next number. Subtraction is done similarly. Multiplication and division, which are somewhat more complicated, call for dividing the abacus into several separate zones, which hold the multiplier or divisor, the multiplicand or dividend, and the answer. Calculations can be carried out past the decimal point to as many places as the size of the abacus allows.

This versatility makes the abacus a remarkably efficient tool—far more so than its predecessors. Until the late Middle Ages, Europeans were stuck with the Roman numeral system, which was almost impossible to use for multiplication or division and so clumsy that it had neither place values nor zero. So for fifteen centuries Europeans did their arithmetical calculations on counting boards that used decimal place values and treated the equivalent of zero as simply a vacant space.

The introduction of Hindu-Arabic numbers, including the essential zero, in Europe in the early thirteenth century led to a bitter controversy. On one side were the "abacists," who used counting boards and recorded the results in Roman numerals. On the other side were the "algorists," who adopted the superior Arabic method and calculated on paper by techniques that the new notation made possible. Today's word algorithm, meaning a step-by-step procedure, comes from the algorists, who in turn took their name from Al-Khowarizmi, a ninth-century mathematician. In some countries, computing by "algorism" was even forbidden by law during the Middle Ages. Not until paper became abundant in the sixteenth century did the new method finally replace the crude Roman system.

Calculating with Arabic numerals on paper or with the help of primitive mechanical devices made of wheels and levers slowly replaced counting boards in Europe. Meanwhile, the abacus became the preferred way of calculating in Russia and Eastern nations. In its contemporary form of beads sliding on rods, it goes back at least to fifteenth-century China and appeared in Japan in the next century.

And so matters remained until the advent of the hand calculator. Today, just as U.S. educators debate how long children should do mathematics drills with pencil and paper before being given access to a calculator, Japanese teachers argue over when to wean their students from the abacus. In both countries, traditionalists argue that unless children first understand the underlying logic of mathematics, they will never understand what a calculator is doing.

But abacus education, which has been part of the grade school curriculum in Japan for more than a century, has now been relegated to ten hours of instruction in the third grade. Even that introduction is enough to intrigue thousands of children with the delights of the soroban. They flock to special schools, called *juku,* for hour-long training sessions three times a week; the cost is around

$10 a month. Some sixty thousand such schools have sprung up all over the country.

National soroban championships are an annual tradition. For the past four years, the hands-down winner has been Eiji Kimura, 21, a senior in business administration at Kyoto Industrial University. His love affair with the abacus began at age eight, when he attended his first *juku.* Two years later he had reached a fourth-degree level (on a scale of ten), and he won his first national championship at age sixteen. Kimura practices with ferocious intensity for two uninterrupted hours every day and can perform feats that flabbergast his rivals and even his teacher. He can add fifteen twelve-digit numbers in twenty seconds and in less than four minutes can perform thirty multiplications, each a twelve-digit number times a six-digit one. Thirty comparable long-division problems take him three minutes.

All this takes place faster than human fingers can flick the beads of an abacus, because Kimura has attained such a state of mathematical *satori* that he now performs his calculations in his head by visualizing the sequences used on a soroban. Here is how

Figure 11.1: A teller at the Mitsubishi Bank in Tokyo does her computing on an electronic calculator but uses an abacus to check the results. Photo by Eiji Miyazawa/Black Star.

he mentally multiplies—256,436 by 1,297,584: "First divide both figures in two parts—256,436 as 256 and 436, and 1,297,584 as 1,297 and 584. Then multiply 436 by 584 and write down the last three numbers. Next multiply 436 by 1,297 and 256 by 584, add them up, and write down the last three numbers. Then multiply 256 by 1,297 and write down the first seven numbers. Now put them in order—the seven numbers first, followed by the three numbers from the second stage of calculation, and finally the three numbers from the first multiplication."

In other words, he immediately breaks multiplication into (256,000 + 436) × (1,297,000 + 584). The whole operation takes him eight seconds. Says he: "I have an abacus in my head, although the image isn't a clear one. When I see a number, I instantly draw mental images of beadlike things."

Division is Kimura's favorite subject, although he concedes that such formidable problems as 3,457,046,665,864 divided by 9,853,796 give him the faintest of headaches: "Sometimes I cannot see the beads clearly in my head. That's when I have trouble. I sometimes see numbers falling apart." Kimura's mentor-coach Masaharu Yamamoto likens his pupil's state of intense concentration in mental calculation to that of Zen meditation—not an ecstatic state but a state of detachment or letting go. Kimura says: "It's not that I don't hear anything. I hear people speaking, but only a few words stick to my memory. Things just don't disturb me."

Many mere mortals, both Asian and Western, experience something of the same soothing feeling while working an abacus. Watching the little beads slide up and down, a child gets an excellent idea of what arithmetic is all about and of how numbers correspond to objects in the real world. Many Western mathematicians and scientists like to use the abacus because of its many sensory charms—the changing visual patterns, the pleasant clicks, the tactile sensation. Perhaps they also enjoy the way the abacus links them to earlier times and other cultures. Perhaps they find a perverse satisfaction in rebelling against the growing, often ugly, complexities of modern life.

Even Japanese calculator manufacturers seem to understand this reluctance to abandon sliding beads for silicon chips. The Sharp Corporation produces an array of calculators that contain small, built-in abacuses. Most owners use the calculator for multiplication and division but are more comfortable doing addition and subtraction on the abacus. Sharp has sold more than 1.5 million of these hybrids in the past decade, a good omen for the survival of a noble tradition.

The Puzzles in Ulysses 12

There are no puzzles in *Ulysses*.
Vladimir Nabokov, *Strong Opinions*

James Joyce was fond of traditional puzzles of all types, both mathematical and linguistic. A footnote on page 284 of *Finnegans Wake* (Viking Compass edition), in a long section of mathematical puns, mentions the last name of Henry Ernest Dudeney, England's most famous creator of mathematical puzzles. In 1905 Joyce was so eager to win a prize of 250 pounds offered by a London magazine for the first correct set of solutions to a puzzle contest that he planned to send his brother Stanislaus a registered, sealed letter of answers in case he later had to prove the date of his entry. Joyce is probably referring to this contest on page 283 of *Ulysses*.[1] Unfortunately his labors went for nothing because his entry arrived too late to qualify.

Joyce's interest in puzzles was almost invisible in *Dubliners,* his book of short stories. It began to surface in his first novel, *A Portrait of the Artist as a Young Man,* and exploded into monstrous proportions in the wordplay and letter play of *Finnegans Wake.* I propose to examine the puzzles that Joyce injected into *Ulysses,* including some enigmas not yet fully solved, and then to consider briefly whether these puzzles add to or detract from the greatness of this greatest of modern comic novels.

Let us begin with letter play. The first letter of the novel *Ulysses* is *S.* In the Random House Modern Library edition (1961) the *S* fills all of page 2. Similarly, the letters *M* and *P* (Pp. 54 and 612, respectively) open the book's second and third parts. Did Joyce intend them to signify something?

This article originally appeared in *Semiotica* 57 (3/4): 317–30 and is reprinted here, with changes and a postscript, with permission.

1. All page numbers refer to the Modern Library edition of 1961. I have checked the just-published three-volume Garland edition, edited by Hans Walter Gabler, which corrects some five thousand typographical errors in earlier publications of Ulysses. None of these corrections influences the material considered here.

As far as I know, Joyce never commented on SMP. It has been observed, however, that in Aristotelian logic, which Joyce studied under Jesuit teachers, *S, M,* and *P* are letters that stand for the three terms of a syllogism: subject, middle term, and predicate. Perhaps Joyce had this in mind. It has also been noticed that the first two words of the book are *Stately, plump.* Their initials provide *S* and *P.* The *M.* is given by Mulligan, to whom both adjectives refer.

Note also that the first section of *Ulysses* opens with *S* and ends with *P.* The third section opens with *P* and ends with *S.* The middle section begins with *M,* and ends with *T.* Molly Bloom's maiden name is Marion Tweedy.

Could *S* stand for Stephen, *M* for Molly, and *P* for Poldy, Leopold Bloom's nickname? These guesses are all in that gray area where no one can be sure of Joyce's intentions, conscious or unconscious, or whether the letter play is coincidental. To make things even grayer, Joyce may have recognized such accidental correlations later and let them stand because he found them amusing.

Poldy is an obviously intended acrostic for the five-line poem (P. 678) that Bloom sent to Molly, but there are other poems in the book that may or may not conceal planned acrostics. Consider the way Joyce breaks the lines of the song (P. 75) Molly plans to sing on her concert tour with her lover Blazes Boylan:

> Love's
> Old
> Sweet
> Song
> Comes lo-ve's old . . .

Did Joyce intend the initial letters to spell *loss,* followed by *comes* to suggest that a loss of love has come to Molly and her husband? The first three lines of a quatrain on page 640 spell *Tao.* This is probably accidental, but who can be sure?

Ulysses swarms with initial-letter abbreviations of which I will cite only a few. KMA and KMRIA (Pp. 146, 147, respectively), presented as newspaper headlines, stand for "Kiss my ass" and "Kiss my royal Irish ass." *Roygbiv* (Pp. 376, 486) are the first letters of the colors of the rainbow. Joyce missed an opportunity to divide them into the name *Roy G. Biv,* but perhaps he deliberately avoided this old mnemonic. (The seven colors are a recurring motif in *Finnegans Wake.*) Many standard Irish abbreviations in the novel are left unexplained. *FOTEI,* for example, stands for Friends of the Emerald Isle, *DMP* for Dublin Metropolitan Police, *DBC* for Dublin Baking Company, and so on. On page 345 there are nineteen such abbreviations in three lines.

A trivial question involving initials was once a topic for speculation by Joyce buffs. On page 237 the names of four men who ride past the College Library on bicycles are mentioned; one is J. A. Jackson. Could the initials J.A.J. be a suggestion that James Augustine Joyce was riding by? Someone checked the Dublin newspapers and discovered that on what is now called Bloomsday (June 16, 1904, the date of the novel) a bicycle race through Dublin actually took place. It was won by J. A. Jackson.

Is it coincidental that *Bella* and *Circe* each have five letters, with their vowels at the same places? We know that Joyce intended *Bella,* the whorehouse madam, to reflect Homer's *Circe,* but whether the correspondence of consonants and vowels was intended remains speculative, though Joyce certainly would have noticed it. There is, of course, no doubt that Joyce intended *Athos,* the name of the dog owned by Bloom's father, to resemble *Argos,* the name of Ulysses' dog in Homer's *Odyssey.*

I know of no high-quality anagrams in *Ulysses.* Bloom's crude attempts to anagram his name (P. 678)—two are flawed by missing a letter—are amusing but scarcely noteworthy.[2] Even the anagrams in *Finnegans Wake* are unremarkable. On page 456, for instance, Joyce scrambles the letters of *steak, peas, onions, bacon, rices, and duckling,* and by substituting *X*'s for consonants and *O*'s for vowels, codes *cabbage* and *boiled Protestants.* This meets his intention of suggesting how chewing rearranges parts of food, but as anagrams they display no unusual cleverness.

Hundreds of words in *Ulysses,* like the tens of thousands in *Finnegans Wake,* are formed by pushing words together in the manner of Lewis Carroll's "portmanteau" words, but most of them are too obvious in meaning to be puzzling. Shakespeare's word *honorificabilitudinitatibus* (from act 5 of *Love's Labour's Lost*), in which consonants and vowels alternate throughout, appears in *Ulysses* on page 210. It is not, however, as long as the 105-letter word (P. 307) which foreshadows the ten great thunderclaps in *Finnegans Wake.*

One of the most outrageous instances of wordplay in *Ulysses* is *AEIOU,* the five vowels in alphabetical order (P. 190). Stephen Dedalus had borrowed a pound from AE., the pen-name of the re-

2. To be fair to Joyce, *Leopold Bloom* is unusually difficult to anagram in a way that makes sense. Dmitri Borgmann, one of the nation's top experts on wordplay, came up with *Loom, bold Pole!* and *Bop Elmo, O doll.* Perhaps the reader can do better.

nowned Irish poet and theosophist George William Russell, which Stephen was supposed to use for food, but instead gave to a prostitute. *AEIOU* is Stephen's way of remembering this debt.

U.P. up is surely the book's most controversial letter play. (See Pp. 158, 160, 280, 299, 320, 381, 446, 474, 486,and 744.) This cryptic message written on an anonymous postcard to the eccentric Mr. Breen arouses him to such fury that he goes to great efforts to find out who sent the card so he can sue for ten thousand pounds. Adams (1962), discloses that *The Freeman's Journal* (November 5, 1903) actually reported a case of one Dublin man suing another for sending a libelous postcard—an incident of which Joyce was probably aware. Molly in her soliloquy (P. 744) recalls her husband "going about in his slippers to look for a £10000 for a postcard up up." Are we to gather from this that Bloom himself sent the card? In any case, what does *U.P. up* mean?

The Oxford English Dictionary (entry U, 2:4) says that when the two letters of *up* are pronounced separately the slang meaning is "over, finished, beyond remedy." A passage is quoted from *Oliver Twist* (chap. 24) where an apothecary's apprentice says "Oh, it's all U.P. there," meaning that he thinks a dying woman will not last more than two hours. Another quotation,"It's all U.P. with him" is explained as "all up either with his health, or circumstances." Adams (1962) calls attention to a line from Arnold Bennett's novel *The Old Wives' Tale* in which a doctor emerges from the room of a dying patient and actually says "U.P. up" Thus the message would be taken by Breen to mean "you'll soon be dead." This interpretation is supported by a mention (P. 474) of *U.P.* as the label of a burial plot. In the French translation of *Ulysses,* authorized by Joyce, the postcard reads *fou tu* (screw you). Change one letter to make *feu tu* and it means "you're dead."

So much for the postcard's primary meaning. Joyce could not, of course, have missed the urinary overtones of *P,* but it is not clear exactly what he had in mind, if anything. Is it to suggest that Breen is impotent—that he can produce only urine, not semen? Adams speculates about several possibilities along such lines, but I think the critics have missed something here. It is not generally known—though, in view of Joyce's (and Bloom's) strong interest in curious aspects of sexual organs, he must have known—that all men bifurcate into two classes. Most men pee down, making it necessary to lift their penis to hit a urinal. A small class pee up, requiring a push down on the penis to get proper aim. As Molly says (P. 743), her husband "knows a lot of mixed up things especially about the body." Was the anonymous writer, perhaps Bloom himself, poking fun at

Breen's membership in the small class of up pee-ers, perhaps with the insulting implication that this is the only way in which Breen's penis is ever up? As Mrs. Breen herself put it (P. 158), the card was sent by "Someone taking a rise out of him."

Joyce was fascinated by the fact that *God,* spelled backward, is dog. This reversal is suggested several times in *Ulysses* and is made explicit in the Black Mass of the Circe chapter. Other reversals occur during the mass, including a backward spelling of *Alleluia, for the Lord God Omnipotent Reigneth* (P. 599). Two old palindromes are quoted (P. 135): Adam's remark to Eve, *Madam, I'm Adam,* and Napoleon's supposed statement, *Able was I Ere I saw Elba.* Reversals of many varieties are almost as common in *Ulysses* as in Carroll's *Through the Looking-Glass.* Mirror reflections are often mentioned, starting with Buck Mulligan's cracked shaving mirror which Stephen sees as a symbol of modern Irish literature. Bloom falls asleep in a curled up position, his "big square feet" so close to Molly's face that she fears he might kick out her teeth (P. 771). In one of the hallucinatory episodes of nighttown, Bloom and Bella exchange sexes. In the role of Bello, the whorehouse madam abuses Bloom unmercifully. Bloom himself has a curious left-right anomaly. We are told (P. 476) that his testicles, instead of hanging in the left trouser like most men's, hang in the right. Perhaps the symmetry in Joyce's own name is worth mentioning: J.A.J. is palindromic, and *James* and *Joyce* each have five letters.

Joyce's fondness for reversals is also indicated by the cipher Bloom uses when he secretly records the name and address of the woman with whom he is having a clandestine correspondence. It is a reverse alphabet cipher: *A = Z, B = Y, C = X,* and so on. The cipher is explained on page 721. *N. IGS./WI.UU. OX/W. OKS. MH/Y. IM* decodes as "Martha Clifford, Dolphin's Barn." Vowels are left out, and the words are taken alternately forward and backward. The periods indicate vowels and slashes divide the four words.

As Kahn (1973, P. 767) points out, the last word, *Barn,* should have been reversed. This could have been Joyce's mistake, but more likely Joyce intended the error to suggest Bloom's carelessness. (The book contains many instances of mistakes by Bloom.) In a similar way Joyce lets us know that Martha, although she responded to Bloom's advertisement for a typist, is a careless typist because in a letter to Bloom (P. 77) she types world when she meant word.

Joyce was (as Kahn informs us) an intimate friend of J. F. Byrne, a man who spent a good part of his life trying to interest governments in a cipher he had invented and which he believed to be unbreakable. The character of Granly in *A Portrait of the Artist* is based on Byrne, and in *Ulysses,* Bloom's address (7 Eccles Street) was Byrne's address in Dublin. Two chapters in Byrne's *Silent Years: An Autobiography with Memories of James Joyce,* deal with his cipher machine. The book contains a message written with this machine and an offer of $5,000 to the first person who decodes it. According to Kahn, the cipher remains unbroken to this day.

I mention all this to suggest how familiar Joyce must have been, through his friendship with Byrne, with cipher systems. Did Joyce incorporate any secret cipher messages (other than Bloom's) in *Ulysses* or (more likely) in *Finnegans Wake?* If so, they have not yet been detected.

Turning from letters to words, *Ulysses* swarms with puns, of which I single out just a few: *Lawn Tennyson* (P. 50), *Lily of the alley* (P. 512), and *met him pike hoses* (a play on *metempsychosis* that recurs throughout the novel). Molly was greatly amused (Pp. 64 and 765) by the name of Paul de Cock, an actual French writer of bawdy novels. *Cuckoo,* taken as a pun on *cuckold* (Pp. 212 and 382), is borrowed from Shakespeare's *Love's Labour's Lost.* Molly's clever pun of *base barreltone* for *base baritone* is recalled by Bloom on page 154 and by Molly on page 759. These and other puns in *Ulysses* are witty but not particularly amusing. *Alfred Lord Tennis Shoes,* for example, is somehow funnier than *Lawn Tennyson.*

Two of the novel's periodic motifs play on the word *throwaway.* On page 151 a YMCA man hands Bloom a throwaway leaflet advertising a lecture by Alexander Dowie, the Scottish evangelist who later founded Zion City, Illinois, a shabby little town north of Chicago where everyone once believed (perhaps some still do) that the earth is flat. Bloom crumples up the leaflet, tosses it into the Liffey, and at intervals we learn of the throwaway's progress as it floats through Dublin.

On page 85 Bantam Lyons asks to see Bloom's newspaper to check the racing page. Bloom tells him to keep the paper because he was going to "throw it away." Bloom is unaware that *Throwaway* is the name of a dark horse (the odds are twenty to one) in the Gold Cup Race that afternoon at Ascot. Thinking he has heard an inside tip, Lyons rushes off to bet on Throwaway. Meanwhile, Blazes Boylan, with whom Molly sleeps that afternoon, has put a bet on a horse named *Sceptre.* He is furious when he learns that Throwaway won the race. As Joyce experts have long recognized, Sceptre is a phallic symbol for the stud Blazes, whereas Bloom is Molly's throwaway husband. However, just as Throwaway wins the race, so Bloom (as we infer from Molly's soliloquy) will probably outlast Blazes in her affections and be the final winner.

"What opera resembles a railroad?" (P. 132). The punning answer is Rose of Castille—rows of cast steel (Pp. 134 and 491 and numerous other pages). There are less interesting riddles in the novel, such as "Where was Moses when the candle went out?" (P. 729), to which the answer (not supplied) is, in the dark. Stephen's riddle about the fox (P. 26) is of no interest to wordplayers. It is a "shaggy dog" riddle—one that cannot be answered unless you know the answer. On the same page Joyce gives the first two lines of an old riddle rhyme

> Riddle me, riddle me, rando ro.
> My father gave me seeds to sow.

Joyce does not supply the next two lines (The seed was black and ground was white./Riddle me that and I'll give you a pipe) or its traditional answer: the speaker is writing a letter.

In *A Portrait of the Artist,* Athy asks Stephen: "Why is the county of Kildare like the leg of a fellow's breeches?" Answer: because there's a thigh in it. After explaining that Athy is a town in Kildare, Athy says there is another way to ask the riddle, but he refuses to tell Stephen what it is. Nor does Joyce tell the reader. This practice of keeping his readers perplexed became an obsession with Joyce, reaching a culmination in *Finnegans Wake.* But even in *Ulysses* there are hundreds of nagging little questions that experts still debate because Joyce has carefully concealed information. We are straying now from wordplay, but a typical example is the problem of whether Bloom's list of twenty-five lovers of Molly (P. 731) gives actual lovers or only men Bloom imagines were lovers. Theories range from the view that Molly actually slept only with Boylan, the last name on the list, to the view that she slept with all of them, including a bootblack at the General Post Office. Joyce obviously wanted his readers to wonder.

Ulysses contains many religious puns, such as those in a blasphemous parody of the Apostle's Creed (P. 329) that begins: "They believe in rod, the scourger almighty." The most amusing religious pun is Bloom's "Come forth, Lazarus! And he came fifth and lost the job" (P. 105). Joyce is here applying an old racing joke about Moses to a passage from the New Testament: "God commanded Moses to come forth, but he slipped on a banana peel and came in fifth."

The quatrain on page 497 seems innocent enough:

> If you see kay
> Tell him he may

> See you in tea
> Tell him from me.

It is the least innocent bit of doggerel in the book. As Joyce experts have pointed out, the first line provides a four-letter obscenity, the third line provides another. Did Joyce also intend the initial letters of the four lines, after moving the last line to the front, to spell *tits?*

Some of the puzzles in *Ulysses* are logical and mathematical. What is the largest number that can be represented by three digits, using no other mathematical symbols? The answer is:

$$9^{9^9}$$

Joyce introduces this old brainteaser on page 699. We are told that when Bloom was trying to square the circle he made a rough calculation of the number's size. He was quite right in concluding that if printed out the number would require some "33 closely printed volumes of 1000 pages each."

Puzzle books are filled with brainteasers about age, such as Sam Loyd's (he was the American counterpart of Dudeney) famous puzzle known as "How old is Ann?" Joyce parodies questions of this sort by speculating at length on the relative ages of Bloom and Stephen (P. 679) if one assumes that, as the years go by, they keep the same ratio their ages had in 1883. As Adams makes clear in *Surface and Symbol,* Joyce's calculations are accurate only for the first dozen lines of this paragraph. In the next line, 714 should be 762, and the numbers 83,300 and 81,396 are also wrong. Is Joyce again letting us know how often Bloom makes mistakes, or (as Adams argues) is it more plausible to assume these errors actually were made by Joyce, who intended the calculations to be correct?

An ancient conundrum concerns a man who points to a picture and says: "Brothers and sisters I have none, yet this man's father is my father's son." Whose picture is it? It is his son's. Joyce modifies this by having Bloom see his "composite asymmetrical image" in a mirror: "Brothers and sisters had he none. Yet that man's father was his grandfather's son" (P. 708). The statement is accurate because Bloom sees himself in the glass. The episode links with Stephen's proof "by algebra" (P. 18) that Shakespeare's grandfather is Hamlet's grandson.

On page 631 a sailor named Murphy displays the number 16 tattooed on his chest but refuses to say what it means. Joyce enthusiasts, worrying over this, have found many references to 16 in the novel. The tattoo is mentioned in the sixteenth chapter of part 3.

The date is June 16. There is an age difference of sixteen years between Bloom and Stephen (P. 679). The Blooms have been married for sixteen years (P. 736). Molly was sixteen when she made her first public appearance as a singer (P. 653). In Europe 16, like 69 in the United States, is a symbol of oral sex.

It is easy to carry this kind of speculation to absurd extremes. The initials of Nora Barnacle (Joyce's wife) are the fourteenth and second letters of the alphabet, and $14 + 2 = 16$. Note also that 16 can be written:

$$2^{2^2}$$

Because Joyce was as secretive as Murphy about the meaning of the tattoo, who can say what, if anything, Joyce intended it to mean?

We do know that all his life Joyce was intrigued by number symbolism and superstitions. He was careful never to make trips or important decisions on the thirteenth day of a month, and he was acutely aware of the fact that his mother died on August 13. When Joyce died on January 13 his wife and friends did not miss the coincidence. Had Joyce been alive in 1955, when his brother Stanislaus died on Bloomsday (June 16), he would not have been surprised.

Joyce often spoke of Dante's obsession with 3, the number of the Trinity. *The Divine Comedy* is in three parts, thirty-three cantos to each, and is written in terza rima. It is probably no accident that *Ulysses* has three parts, with three chapters to the first and final parts, and $3 \times 4 = 12$ chapters in the middle part. Molly's age on Bloomsday is thirty-three (P. 751). In conversation with Adolph Hoffmeister, Joyce once discussed the significance of 12—the twelve apostles, twelve tables of Moses, twelve months, and so on. "Why," Joyce asked, "was the armistice of the Great War trumpeted forth on the eleventh minute of the eleventh hour of the eleventh day of the eleventh month?"

We conclude with some unclassified puzzles. After the famous episode in which Bloom, hands in pockets, masturbates while Gerty allows him glimpses of her underdrawers, Bloom picks up a stick and, like Jesus, writes in the sand (P. 381). He writes "I AM A" but never completes it. How did he intend to finish? Man, Jew, fool, cuckold, masturbator? Or is it supposed to mean "I am alpha" the beginning of all things? There is no consensus among scholars.

On page 761 of the flawed Modern Library edition of Ulysses, before it was corrected in 1961, an unintended period slipped into Molly's unpunctuated soliloquy. Molly's actual period starts on page

769, forcing her out of bed and onto a cracked chamber pot. And when was Molly's birthday? We learn on page 736 that it was September 8, the traditional date for celebrating the birth of the Virgin Mary. There are more 8's connected with Molly. Her marriage, at eighteen, was on October 8, 1888 (P. 736), and her monologue, perhaps not by accident, consists of eight sentences. It has also not escaped Joyceans that when 8 is rotated 90 degrees it becomes the symbol for infinity and eternity.

The greatest of all unresolved puzzles in Ulysses is the identity of a "lankylooking galoot" in a brown mackintosh who first turns up at the funeral Bloom attends in the Hades chapter. No one there knows who he is, and throughout the novel Bloom wonders about him. We are told that he was the thirteenth mourner—"death's number" Bloom says to himself—and it may be intentional that there are thirteen references to him in the book (109–110, 112, 254, 290, 333, 376, 427, 485, 502, 511, 525, 647–648, and 729). We learn that he "loves a lady who is dead." A newspaper account of the funeral calls him McIntosh, but that is a mistake. The reporter heard Bloom use the word mackintosh and wrongly took it to be the man's name.

We glimpse the mystery man on the street, "eating dry bread" (P. 254) and later encounter him in a bar, near the redlight area, where he is drinking Bovril soup (P. 427). (Bovril is the trade name of an instant beef soup introduced in England in 1889. It was widely advertised with the slogan, appropriate to its context here, "Bovril prevents that sinking feeling.") The man's seedy clothes are described, and we learn that he was once a prosperous citizen "all tattered and torn that married a maiden all forlorn. Slung her hook, she did. Here see lost love." The man "thought he had a deposit of lead in his penis. Trumpery insanity. Bartle the Bread we calls him. . . . Walking Mackintosh of lonely canyon." I have no idea what Bartle means.

In the whorehouse dream sequence the man in the brown raincoat pops onto the stage through a trap door, points a finger at Bloom, and says "Don't you believe a word he says. That man is Leopold M'Intosh, the notorious fireraiser. His real name is Higgins" (P. 485). Ellen Higgins was Bloom's mother, and one of the prostitutes is Zoe Higgins. What Joyce is trying to say here is far from clear.

"Who was M'Intosh?" Joyce asks explicitly (P. 729). We are told in Gorman's (1948) biography of Joyce that Joyce was fond of asking friends this question, but he always refused to answer. There have been many theories. Among the implausible:

1. Theoclymenus, a Greek soothsayer whose presence in Homer's *Odyssey* (Books 15 and 20) is somewhat mysterious;
2. Wetherup, an obscure acquaintance of Joyce who is mentioned in *Ulysses* (Pp. 126 and 660);
3. James Duffy, a character in "A Painful Case" (*Dubliners*) based in part on Joyce's brother Stanislaus;
4. The ghost of Charles Parnell, the Irish nationalist leader;
5. The Wandering Jew;
6. Jesus;
7. Nobody (Joyce was playing a joke on his readers).

Hodgart (1978) takes the mystery man to be Death. Vladimir Nabokov (1980) argues that the man in the brown raincoat is none other than Joyce himself, a "selfinvolved enigma" (P. 729) that Bloom could not comprehend. Joyce certainly was a "lankylooking galoot." The man's lost love could be the Virgin Mary, symbol of the Roman Catholic Church which had its hooks into Joyce before he abandoned her. Perhaps the bread and soup are intended to contrast with the bread and wine of the Eucharist. The conjecture may be further supported by the quotation ("all tattered and torn . . .") from the Mother Goose rhyme "The House That Jack Built." Joyce grew up in a household that Jack (his father John) built.

After the word *Bovril* (P. 427) Joyce adds "by James". Bovril was introduced in England by someone named J. Lawson Johnston (according to the Oxford English Dictionary Supplement), but whether the J stands for James, I have been unable to determine. Is it possible Joyce wants us to take this to mean that the soup is on the table next to James Joyce? In any case, Nabokov is convinced that Joyce, as Stephen insists Shakespeare often did in his plays, "set his face in a dark corner of his canvas" (P. 209). Perhaps when Bloom hears the loud crack in a table and turns out the light (P. 729) he suddenly comprehends the awful truth—he is no more than a figment in the mind of a writer who has imagined him.

We know Joyce is not the man who races through Dublin on a bicycle, and he may not be the man in the brown raincoat, but there is little doubt that Joyce puts in a totally unexpected appearance in Molly's monologue, "O Jamesy," she cries out when she realizes she is menstruating and about to stain the bed's clean sheets, "let me up out of this pooh sweets of sin" (P. 769). *Sweets of Sin* is the title of a trashy novel that Bloom has brought home for his wife. Who could Jamesy be but Molly's creator? Like many another novelist before and since, Joyce could not resist this moment of paradoxical self-reference—an imaginary woman calling out to the man who imagined her and who wrote the very words she is using.

Now we must ask the question, How much does all this riddling add to the worth of *Ulysses?* The wordplay attributed to Bloom does indeed enrich our understanding of Bloom. It lets us know that he, like Homer's Ulysses, is a man of many wiles. He likes anagrams and acrostics. He knows enough geometry to attempt squaring the circle. We are given a catalogue of his many ingenious schemes for making money. He uses a cipher. But what about the wealth of wordplay that is not Bloom's but Joyce's?

Numerous writers have enjoyed peppering their fiction with outlandish wordplay. One thinks of the puns in the works of Aristophanes and Rabelais and Shakespeare. Even Milton had a weakness for puns. In American literature one thinks of the incessant wordplay of James Branch Cabell, Peter DeVries, and Vladimir Nabokov. Sometimes the play turns up where least anticipated. In F. Scott Fitzgerald's *This Side of Paradise,* for example, all the purple passages in italics are formal poems, not very good ones, disguised as prose. In chapter 7 of *The Great Gatsby* we learn of a man named "Blocks" Biloxi who makes boxes and comes from Biloxi, Mississippi, but what relevance this has to the story is not apparent. Even the novel's title, like that of Finnegans Wake, conceals a pun. Fitzgerald well knew that gat was then underworld slang for a handgun. Readers who relish wordplay, coming upon such spots in a novel, may be as pleased as the author was when he put them there, but do such whimsies improve the novel?

My own view is a dull compromise. As a longtime connoisseur of puzzles, I am not offended by Joyce's riddling. On the other hand, neither am I much impressed. The sad truth is that the wordplay in Ulysses is not on the highest level. It takes only a glance at books such as Dmitri Borgmann's *Language on Vacation* or at the pages of *Word Ways* (an American quarterly devoted to recreational linguistics) to realize how trivial most of it is. Any clever writer can compose acrostics, toss in old riddles, blend two or more words into one, concoct puns, and hide meanings under thick layers of enigmatic persiflage. No skill is needed to spell a sentence backward or to observe that *dog* is a reversal of *God.* Joyce simply was not capable of inventing a palindrome comparable to, say, "Straw? No, too stupid a fad. I put soot on warts."[3] The wordplay in Ulysses may

---

3. I selected this palindrome from hundreds that may be found in Bergerson (1973). A semordnilap is a sentence that spells a different sentence when taken back-

indeed add to the novel's overall comic atmosphere, but in my opinion it does not add much.

We can focus the issue by considering two cryptic words that appear side by side on page 286: *yrfmstbyes* and *blmstup.* There is no doubt that the second word means "Bloom stood up" because Joyce himself explains it two sentences later. Even if we grant that *blmstup* reinforces a mental image of Bloom suddenly standing up, is it worth the effort of going back to decode it? And what on earth does *yrfmstbyes* mean?

To put the words in context, Bloom has just finished dining in a hotel restaurant with Stephen's alcoholic uncle. He thinks: "Well, I must be. Are you off? *Yrfmstbyes. Blmstup.*" In the vast literature on *Ulysses* I am sure there must be theories about *yrfmstbyes,* but I have not encountered them, so let me pass along two suggestions. Borgmann thinks that Bloom is answering "Well, I must be. Are you off?" with "You are off. Must be, yes." My wife came up with a more startling conjecture. From his seat in the dining room Bloom has been watching a barmaid in the adjoining saloon. She is massaging a beerpull knob by sliding her fingers smoothly back and forth over the "firm white enamel baton" while Bloom is mentally masturbating. Could he be answering "Are you off?" with "You royal fucking masturbator, yes"? Perhaps Joyce slyly contrived the preceding words so that both interpretations could be made.

For readers who like to solve cryptograms, *yrfmstbyes* and *blmstup* may add interest to the texture of a novel. For readers who care little about such conundrums, the strange words are mere blots on the text. *Bltsnthtxt.*

As for *Finnegans Wake,* I agree with Nabokov. Here Joyce's preoccupation with wordplay, kept under control in *Ulysses,* overwhelms everything else. Searching for the plot, philosophy, and beauty below the surface of what Nabokov called *Punnigans Wake* may forever Joyceously occupy erudite and multilingual critics, but I suspect the world's final verdict will be that the book is little more than a monstrous linguistic curiosity. Even the vast knowledge of world literature, the high intelligence, the stylistic skill, the humor, and the tireless energy that went into the making of this mammoth dish of verbal "chop suey" (as Joyce's wife called it) become less awesome when you consider that Joyce had sixteen years (16 again!) to cook it. In an interview Nabokov said: *Ulysses* towers over

---

ward letter by letter. An example, supplied by Borgmann, is *Rail at natal bosh, aloof gibbons.* Properly punctuated, it reads in reverse: *Snob! Big fool! Ah, so blatant a liar!*

the rest of Joyce's writings, and in comparison to its noble original-
ity and unique lucidity of thought and style the unfortunate *Finnegans Wake* is nothing but a formless and dull mass of phony folklore, a cold pudding of a book, a persistent snore in the next room,
most aggravating to the insomniac I am. . . . *Finnegans Wake*'s fa-
cade disguises a very conventional and drab tenement house, and
only the infrequent snatches of heavenly intonations redeem it
from utter insipidity (Appel 1967, Pp. 134–35).

References

Adams, Robert M. *Surface and Symbol: The Consistency of James
   Joyce's Ulysses.* New York: Oxford University Press, 1962.
Appel, Alfred Jr. Interview with Vladimir Nabokov. *Wisconsin
   Studies in Contemporary Literature* 8 (Spring, 1967) 134–
   135.
Bergerson, Howard W. *Palindromes and Anagrams.* New York:
   Dover Publications, 1973.
Ellmann, Richard. *James Joyce.* New York: Oxford University
   Press, 1959.
Gifford, Don and Seidman, Robert J. *Notes for Joyce: An Annota-
   tion of James Joyce's* Ulysses. New York: R.P. Dutton, 1974.
Gorman, Herbert. *James Joyce.* New York: Rinehart, 1948.
Hodgart, Matthew. *James Joyce: A Student's Guide.* London: Rout-
   ledge and Kegan Paul, 1978
Kahn, David. *The Codebreakers.* New York: New American Library,
   1973.
Nabokov, Vladimir. *Lectures in Literature.* New York: Harcourt
   Brace Jovanovich, 1980.

Postscript

A puzzle I failed to mention occurs on page 581 in the brothel episode. Stephen says: "Tell me the word, mother, if you know now.
The word known to all men."

Stephen's mother does not supply the word, and critics have
disagreed on what it could be. Richard Ellmann, in *Ulysses on the
Liffey* (1972), suggested *love.* Another critic thought it *death* and
still another proposed *synteresis.* This would seem, commented Ellmann (writing on "The Big Word in Ulysses," *New York Review of*

Books, October 25, 1984), "rather to be the one word unknown to all men."

The mystery was solved in 1984 when the corrected three-volume edition of *Ulysses* was published. A passage had been omitted from the Scylla and Charybdis episode in which Stephen says: "Do you know what you are talking about? Love, yes. Word known to all men." This is followed by a Latin quotation which Ellmann translates as "Love truly wishes some good to another, and therefore we all desire it." It is not known whether Joyce wanted the passage removed or whether it was inadvertently dropped.

On page 153 Bloom reads a sign that says POST NO BILLS. This is followed by POST 110 PILLS—puzzling unless you realize that someone has scraped off the diagonal line in *N* and the bottom loop of *B*. Someone also called my attention to the made-up palindromic word *TATTARRATTAT,* though I have not yet located it in the novel.

Everett Bleiler was intrigued by the difficulty of anagramming Leopold Bloom. "Do you think it possible," he asked in a letter "that Joyce left an incident out of *Ulysses?* When Bloom, having robbed the Dublin post office, went with the loot to his gangster mistress, she wanted to make love, but he insisted instead on her giving him a haircut. When her scissors accidentally cut his ear, they quarreled and she kicked him out. I don't remember what happened to the loot."

All this is necessary background for understanding the following dialogue, each line of which is an anagram on Bloom's name!

> "Bloom! Ope, doll!
> P.O. boodle, moll!"
> "Bold pool-mole,
> O, do loll, bop me!"
> "P.L.O.-model bolo?
> Poll me. O! Blood!
> Doombell! Loop!"
> "O!! Do lollop, B.E.M.!
> Plod, Leo Bloom!"
> Lo, lo, bold poem!

13

The Fantasies of H. G. Wells

A secular humanist, social critic, novelist, H. G. Wells produced voluminous works that reflected a lifelong passion for combating injustice and building a better world. Reason and science were for Wells the tools by which the human race, as it slowly and painfully emerges from its bestial past, is constructing a world culture free of superstition, war, poverty, and disease. He called this movement the "open conspiracy" to distinguish it from the closed, clandestine conspiracies of radical movements inspired by Karl Marx. *Men Like Gods* (1923), the greatest of his utopian novels, uses the device of a "parallel world" to paint a glowing picture of what earth's future could be like. But Wells also had dark moods in which he described negative utopias that might result should the conspiracy fail, especially if progress in modern weaponry causes mankind to lose what Wells called, in an often quoted phrase, the "race between education and catastrophe."

Herbert George Wells (1866–1946) was born on September 21 at Bromley, Kent, the son of a shopkeeper and semiprofessional cricket player and a woman who had been a lady's maid. After obtaining a degree in biology, he taught for a short while before starting his distinguished writing career. His first big success, *The Time Machine* (1895), became a classic of science fiction. Although Wells was greatly admired for his realistic novels, such as *Kipps: The Story of a Simple Soul* (1905) and *Tono-Bungay* (1909), and for his monumental trilogy of knowledge, *The Outline of History* (1920), *The Science of Life* (1930), and *The Work, Wealth and Happiness of Mankind* (1931), he is read today mainly for his science fantasies.

This article originally appeared in *Supernatural Fiction Writers: Fantasy and Horror,* vol. 1, edited by E. F. Bleiler, and is reprinted here with the publisher's permission. ©1985 by Charles Scribner's Sons.

Wells's reputation as the father of modern science fiction rests not only on the quantity and quality of his work in this genre, but also on the astonishing number of basic science-fiction devices that he was the first to use in notable ways. In almost all his fiction, including his science fiction and fantasy, Wells was concerned with more than entertainment. His novels and short stories usually contain philosophical or political morals, often in the form of harsh satire on social customs and institutions toward which he was unsympathetic.

Wells wrote only five short novels that can be called fantasy, though of course fantasy pervades his science fiction. *The Wonderful Visit* (1895) was suggested by a remark of John Ruskin's to the effect that if an angel appeared on earth, someone would be sure to shoot it. The novel opens in a small London suburb where a vicar shoots what he thinks is a flamingo. It turns out to be a beautiful male angel, not from the Christian heaven but from a world in hyperspace where there is no evil, sickness, or growing old. While the liberal-minded vicar nurses the immortal back to health, the petty, ugly reactions of people in the community provide Wells with grist for attacking British culture and contrasting it to the utopian socialist vision symbolized by the world from which the angel came.

The angel's gentle, clumsy attempts to understand and adapt to human society create many problems, not least of which is his romance with Delia, the vicar's pretty housemaid. Although the angel finds it difficult to eat with knife and fork and to sleep on a bed, he proves to be an accomplished violinist. The music he produces arouses in the vicar a vision of such strange and supernal beauty that the vicar vows never to play his violin again.

One night the vicar, after lighting his reading lamp, carelessly drops the unextinguished match into a wastepaper basket. The vicarage catches fire. Delia rushes into the flames to save the vicar's violin, and the angel follows. Both are translated to the other world after perishing in the fire. The vicar, awakened by the angel to the world's stupidities, dies soon thereafter. Wells collaborated with St. John Ervine on a stage version of *The Wonderful Visit* for a London production at St. Martin's Theatre in 1921. Today the novel is almost completely forgotten.

Equally unremembered is Wells's longest fantasy novel, *The Sea Lady: A Tissue of Moonshine* (1902). It tells the story of another immortal, not from the sky but from the sea—a golden-haired mermaid who comes out of the ocean at Sandgate beach to investigate human life, especially the life of Harry Charteris, to whom she had been attracted. Charteris has before him the promise of a conventional marriage and a career in politics. Nevertheless, he finds himself sensuously drawn toward the sea lady and her whispers of "better dreams" and a region of mystery that transcends the world he knows. In Wells's previous portrayal of a "wonderful visit" the vicar conceals the angel's wings by giving him a large overcoat to wear. Charteris disguises his "angel" as Miss Doris Thalassia Waters, whose fish tail is always covered when she is taken to the beach in a wheelchair.

Miss Waters is, of course, a symbol—the female counterpart of the seaman to whom Henrik Ibsen's heroine is attracted in his play *The Lady from the Sea*. In both works the theme is the conflict between a safe, dull, predictable life and the wild, lawless dreams, the siren songs, of sexual love and adventure. It is a conflict Wells explored later in *The New Machiavelli* (1911) and several other realistic novels. *The Sea Lady* ends when Charteris, carrying the mermaid in his arms, walks into the glittering moonlit sea, "hastening downward out of this life of ours to unknown and inconceivable things."

Although Wells's novel *The Undying Fire* (1919) is modeled on the Book of Job, opening with a prologue in which God and Satan argue about good and evil and the future of humanity, it is essentially a realistic novel—the story of an educator dedicated to the "undying fire" of knowledge that older generations must pass on to younger ones. Considered separately, however, the book's prologue is a gem of philosophical fantasy.

Three of Wells's short novels, each published as a small book, may be called fantasies. *The Croquet Player* (1936) has the familiar Wellsian theme that man is still an animal, quite capable of irrational self-destruction. The story is told by George Frobisher, an indolent, conservative, upper-class Englishman who has no interest in the "open conspiracy." A retired doctor half persuades him that a nearby area called Cainsmarsh is haunted by the ghosts of Neanderthal cavemen. In a sense the story is not fantasy, because a psychiatrist, whose views are those of Wells, enters and tells Frobisher that Cainsmarsh does not exist. It is a delusion fabricated by the doctor's unconscious mind so that he can cope with his realization that the world is going mad. The world is indeed on the brink of destruction (Wells was writing in the shadows of the coming Second World War). "Only giants can save the world," the psychiatrist shouts at Frobisher. "We have to bind a harder, stronger civilization like steel about the world. We have to make such a mental effort as the stars have never witnessed yet. Arise, O Mind of Man!"

This Wellsian rhetoric falls on uncomprehending ears. Says

Frobisher: "I don't care. The world *may* be going to pieces. The Stone Age may be returning. This may, as you say, be the sunset of civilization. I'm sorry, but I can't help this morning. I have other engagements. All the same—laws of the Medes and Persians—I am going to play croquet with my aunt at half-past twelve today."

The *Camford Visitation* (1937) is a brief, inconsequential attack on England's higher education. Its fantasy element centers on another "wonderful visit"—this time by a being from a higher space-time who has been observing life on earth for millions of years in the way an earthling, out of curiosity, might observe an anthill. The being is never seen. Only its voice is manifest—an inhuman, metallic voice that is heard by leaders of the university town of Camford (the name conflates the names of *Cam*bridge and O*xford*). The voice warns the university community of the world's impending suicide and urges a great educational effort to avert it; but those who hear the voice are as indifferent to the warning as Wells's croquet player.

All Aboard for Ararat (1940) sounds the same note of doom. Noah Lammock is a writer who, like Wells, has tried vainly to arouse the mind of man. To his house, after having escaped from a mental institution, comes none other than Jehovah himself, with white woolly hair and long beard, to tell Noah he must build another ark. In addition to a selection of animals and a crew of admirable men and women, he is to carry the essentials of world knowledge on microfilm. Jehovah is not very intelligent or well informed, and there is much amusing dialogue between Noah and God that allows Wells to take sharp jabs at Old Testament mythology. The story breaks off abruptly. When it resumes Noah has been piloting the ark for more than a year, waiting for the waters to subside while he searches for the top of Ararat. God is a low-ranking member of the crew, with the tasks of preaching on Sundays and playing the harmonium. The future of mankind is uncertain.

None of the fantasies mentioned approach the excellence of Wells's major science-fiction novels: *The Island of Dr. Moreau* (1896), *The War of the Worlds* (1898), and *The First Men in the Moon* (1901). It is only in some dozen short stories that Wells produced memorable fantasy. With one exception, all these tales can be found in the large anthology *The Short Stories of H. G. Wells* (1927).

"The Man Who Could Work Miracles" is the best known of these fantasies and is the only one that became a full-length motion picture. Wells himself wrote the script for the 1936 London-made film starring Roland Young. The script was published that same year as a book and later was reprinted in *Two Film Stories* (1940) together with Wells's scenario for *Things to Come,* the most successful of many motion-picture adaptations of his fiction.

George Fotheringay, the man who could work miracles, is a meek clerk who, while arguing in a London pub that miracles cannot occur, discovers that whatever he commands happens. After some trivial miracles, the clerk begins to experiment with more-grandiose ones. To test the extent of his miraculous power he commands the earth to stop rotating, not anticipating the monstrous inertial effects that would result. Centrifugal force propels all objects on the planet, including Fotheringay, into space. He quickly wills himself safe on the ground and in the roaring hurricane asks that history return to that moment in the pub when he first discovered his inexplicable power, but now he requests that the power be denied him. The tale instantly shifts back in time, with all memory of what had occurred erased from the clerk's mind. The moral is evident: if natural law could be suspended by true miracles, the results would be catastrophic.

Two of Wells's stories that lean more toward fantasy than toward realism or conventional science fiction are based on orthodox Christianity, which Wells, of course, did not take seriously. In "A Vision of Judgment" certain sinners, on Judgment Day, flee in shame up God's sleeve after the Recording Angel reads aloud a summary of their lives. They are given a chance to try again when God shakes them out of his sleeve, with new bodies, onto a planet that orbits Sirius. When the story first appeared in *The Butterfly* (September 1899), it was illustrated by S. H. Sime, who later became famous as the illustrator of Lord Dunsany's many fantasies.

In "The Story of the Last Trump" a child playing in an attic in heaven finds the huge brass trumpet reserved for Judgment Day. He drops it to earth, where it turns up in a pawnshop and is bought by two men. They are unable to sound it until they apply a powerful bellows. For a microsecond, all over the earth, God and the angels are seen in the sky; then a hand of fire reaches down to snatch the trumpet. The world returns to normal.

Also concerned with Christian mythology, although associated with two other stories about doubt and nonacceptance, is "The Apple." A stranger gives a golden apple to a young student on his way by train to London University. The stranger insists it is an apple from the Tree of Knowledge but that he has lacked the courage to eat it. The skeptical student tosses the fruit away. In a dream he realizes it was indeed the forbidden fruit, but his efforts to find it again are in vain.

In "The Temptation of Harringay" a struggling artist watches a figure that he is painting on a canvas come to life. It is a devil, and it offers him the ability to paint masterpieces in exchange for his soul. Harringay obliterates the face with green enamel. Since then he has never produced a great painting.

A cabinet minister, in "The Door in the Wall," is haunted by a green door in a white wall. As a child he had once walked through it into a Garden of Eden—an enchanted utopia of beautiful people. At intervals in his life the door mysteriously appears, but each time something prevents him from entering. One day he is found dead at the bottom of an excavation. He had walked through a door carelessly left unfastened in a protective fence. The story, a parable of paradise lost, is echoed by Thomas Wolfe in his haunting refrain, "A stone, a leaf, an unfound door." Wells's tale was made into a short British film in 1956 by using a special screen that expanded, contracted, and masked off portions of the picture for special effects.

More closely related to the contemporary story of supernatural horror, albeit psychological rather than physical, is "The Red Room." A man spends the night in a haunted room of a decaying castle. No ghosts appear, but a strong atmosphere of evil and foreboding and his inability to keep candles from going out make him flee in terror.

Several of Wells's supernatural short stories concern themselves with psychic research and the problems of ghosts and/or spirits. In "The Inexperienced Ghost" Clayton encounters the phantom of a weak, ineffectual young man. Uncertain of what he is supposed to do, the phantom has been trying desperately to haunt the golfing club where Clayton is spending the night. Clayton helps the ghost recall the gestures necessary for transporting himself back to the other world. When Clayton tells his friends about the incident, they refuse to believe him. As an experiment Clayton repeats the mystic hand passes made by the ghost and instantly drops dead.

In "The Stolen Body" Mr. Bessel, a businessman interested in psychic research, projects his soul from his body. The soul floats about over London in a shadowy hyperspace filled with mute drifting spirits of the dead. Meanwhile, an evil soul takes possession of his body. It runs wildly through London streets screaming "Life! Life!" and smashing people with a cane. After the stolen body falls down a shaft on Baker Street, where it lies battered, the evil soul abandons it and Bessel is able to enter it again.

Similarly straddling the line between fantasy and psychological science fiction, "The Story of the Late Mr. Elvesham" tells how an aged philosopher, Egbert Elvesham, manages (with the aid of mysterious chemicals) to swap bodies with a young student and thus evade his mortality. Eden, the student in Elvesham's body, commits suicide. Elvesham, in Eden's body, is killed by a London cab.

Magic and fairy lore appear in two stories. In "The Magic Shop" a father and small son wander into a conjuring shop in London where they are entertained by the owner, who insists that his magic is genuine. When the pair find themselves back on Regent Street, the shop has vanished. The boy later tells his father that the toy soldiers bought in the shop come alive whenever he says a certain secret word. In "Mr. Skelmersdale in Fairyland" a handsome young grocery clerk asleep on an enchanted knoll wakes to find himself in fairyland. He falls in love with the fairy lady who had brought him there because of her passion for him. When he insists he must go back to Millie, the girl to whom he is engaged, the fairy lady sends him home. The gold that gnomes had stuffed into his pockets has turned to ashes. Although he longs for his lost love and tries desperately to return to the fairy world, he is unable to fall asleep again on the knoll.

"A Dream of Armageddon" is similar in background to Wells's science-fiction works "A Story of the Days to Come" and *When the Sleeper Wakes* (1899). In recurring dreams a Liverpool solicitor lives another life at an unspecified time in the future. In the dream he is a powerful leader who has abandoned British politics to spend the rest of his life abroad with the woman he loves (the *Sea Lady* theme). An evil rival has become the head of a fascist movement that threatens world conflict. The dreamer is torn between a desire to return to England to defeat the movement and a desire to stay in Italy with his beloved. He chooses to stay. Both are killed in the inevitable war.

Dream is also the subject of "Under the Knife," in which a man, put to sleep with chloroform, dreams he is killed by the surgical operation he is undergoing. Before he wakes to learn otherwise, his soul leaves the solar system, expanding until the entire universe shrinks to a glittering speck on the ring worn by a vast hand. Our suns are atoms in a larger universe, perhaps in turn the atoms of a still larger one, and so on into an infinite regress.

In "Answer to Prayer," in a moment of agony a liberal bishop prays for help. When a voice answers, "Yes. What is it?" he dies of fright. Published in British and American periodicals in 1937, this is one of several short stories by Wells that are not in any of his book collections.

Wells, who died on August 13, 1946, lived to read about the

destruction of two Japanese cities by the atom bomb, which he had foreseen and named in his prophetic novel *The World Set Free* (1914). His last two books, *The Happy Turning* (1945) and *Mind at the End of Its Tether* (1945), are brief expressions of the two moods that alternated throughout Wells's life. The first of these small books relates a dream in which Wells makes a happy turn that allows him to walk into the fields of Elysium the way the cabinet minister in Wells's youthful story walked into them through the green door. Wells imagines remarkable conversations with Jesus, who considers his life a failure and has nothing but contempt for Christianity. The dream's mood is one of optimism for the future of mankind. The other book is an expression of profound despair, a statement of Wells's fear that nothing now can save humanity from self-obliteration.

SELECTED BIBLIOGRAPHY

Fantasy Works of H. G. Wells

The Wonderful Visit. London: Dent, 1895, New York: Macmillan, 1895. (Short novel)
The Sea Lady: A Tissue of Moonshine. London: Methuen, 1902. Westport, Conn.: Hyperion, 1976. (Novel)
The Short Stories of H. G. Wells. London: Ernest Benn, 1927. Garden City, N.Y.: Doubleday, 1929. (Sometimes reprinted under the title *The Complete Short Stories of H. G. Wells)*
The Croquet Player. London: Chatto and Windus, 1936. New York: Viking, 1937. (Short novel)
The Man Who Could Work Miracles. London: Cresset Press, 1936. New York: Macmillan, 1936. (Wells's film script for a motion picture based on the short story)
"Answer to Prayer." *New Statesman and Nation* (April 10, 1937). *The New Yorker* (May 1, 1937). (Short story)
The Camford Visitation. London: Methuen, 1937. (Short novel)
All Aboard for Ararat. London: Secker and Warburg, 1940. New York: Alliance Book Corporation, 1941. (Short novel)

Critical and Biographical Studies

Bergonzi, Bernard. The Early H. G. Wells: *A Study of the Scientific Romances*. Toronto: University of Toronto Press, 1962.
Brome, Vincent. *H. G. Wells: A Biography*. London: Longmans, Green, 1951.

Haining, Peter, ed. *The H. G. Wells Scrapbook.* New York: Clarkson Potter, 1979.

Huntington, John. *The Logic of Fantasy: H. G. Wells and Science Fiction.* New York: Columbia University Press, 1982.

McConnell, Frank. *The Science Fiction of H. G. Wells.* Oxford: Oxford University Press, 1981.

MacKenzie, Norman, and MacKenzie, Jeanne. *The Time Traveller: The Life of H. G. Wells.* London: Weidenfeld and Nicolson, 1973. As *H. G. Wells: A Biography.* New York: Simon and Schuster, 1973.

Ray, Gordon N. *H. G. Wells and Rebecca West.* New Haven, Conn.: Yale University Press, 1974.

Vallentin, Antonia. *H. G. Wells: Prophet of Our Day.* New York: John Day, 1950.

Wagar, W. Warren. *H. G. Wells and the World State.* New Haven, Conn.: Yale University Press, 1961.

Wells, H. G. *Experiment in Autobiography. Discoveries and Conclusions of a Very Ordinary Brain—Since 1866.* London: Gollancz and Cresset Press, 1934. New York: Macmillan, 1934.

West, Geoffrey [Wells, Geoffrey H.] *H. G. Wells: A Sketch for a Portrait.* London: Howe, 1930. New York: Norton, 1930.

BIBLIOGRAPHIES

Hammond, J. R. *Herbert George Wells: An Annotated Bibliography of His Works.* New York: Garland, 1977.

H. G. Wells Society. *H. G. Wells: A Comprehensive Bibliography.* London: H. G. Wells Society, 1966.

Wells, Geoffrey H. *The Works of H. G. Wells, 1887–1925. A Bibliography, Dictionary, and Subject-Index.* London: Routledge, 1926.

The Fantasies of G. K. Chesterton

14

Fantasy, G. K. Chesterton never tired of saying, should remind us of how fantastic the real world is. No imaginary animal could be more unlikely, he wrote, than a rhinoceros or a pelican—or a human animal balanced on its hind legs and obtaining energy by pushing food and pouring liquids through a hole in its head. Over and over again, in his fiction, nonfiction, plays, and poetry, Chesterton stressed the bizarre and miraculous aspects of ordinary things. The notion that fairy tales are unhealthy for children seemed to him close to mortal sin; his essay "The Dragon's Grandmother" (in *Tremendous Trifles,* 1909) was his most entertaining blast at what he thought was an absurd contention. In numerous other books he insisted that in some respects fairy tales are truer to life than is realistic fiction.

Gilbert Keith Chesterton, or G. K., as he came to be known, was born in London on May 29, 1874, and died on June 14, 1936. He left school at seventeen to study commercial art, and although he never tired of sketching, his major talent lay in writing. His first novel, *The Napoleon of Notting Hill* (1904), is a wild, improbable tale, set in 1984, the year destined to be made famous by George Orwell. But it is not fantasy in the usual sense . Almost all Chesterton's other novels and short stories are equally unrealistic in their grotesque, unbelievable plots, but when the supernatural is not explicitly invoked, it would be improper to call them fantasies.

Chesterton's first and greatest fantasy, *The Man Who Was Thursday* (1908), was published fifteen years before he left the Anglican church of his parents to become a Roman Catholic. It is a novel about a Catholic poet, Gabriel Syme, who is given the name "Thursday" when he joins a London group of anarchists plotting the

This article originally appeared in *Supernatural Fiction Writers: Fantasy and Horror,* vol. 1, edited by E. F Bleiler, and is reprinted here with the publisher's permission. © 1985 by Charles Scribner's Sons.

destruction of civilization. Each of the seven members of the Central Anarchist Council is known by a day of the week. Sunday, the council's rotund leader, slowly emerges as a monstrous symbol of nature—god in pantheistic immanence rather than biblical transcendence.

Sunday has the face of an archangel, but from the back his huge bulk resembles a beast. This double aspect of nature is reinforced by the gradual disclosure that all the other council members, including Thursday, are spies hired by a mysterious officer at Scotland Yard who is later revealed to be Sunday himself. For some unfathomable reason Sunday, seemingly indifferent to good and evil, is orchestrating both sides. There is a mad chase after Sunday, who keeps tossing nonsense notes to his pursuers while he flees, first in a hansom cab, then on the back of an elephant, and finally in a balloon. After a fantastic costume ball, Syme realizes that anarchy is a necessary consequence of god's gift of free will and the right to suffer. He confronts Sunday with the question, "Have you ever suffered?" "As he gazed, the great face grew to an awful size, grew larger than the colossal mask of Memnon, which had made him scream as a child. It grew larger and larger, filling the whole sky; then everything went black. Only in the blackness before it entirely destroyed his brain he seemed to hear a distant voice saying a commonplace text that he had heard somewhere, 'Can ye drink of the cup that I drink of'"? (Chap. 15).

This is the only hint in the book that nature is a mask worn by the Christian god. Syme regains consciousness to discover that his adventures were part of a long nightmare. Exactly what the dream signifies has been much debated. Chesterton himself struggled to explain it in his introduction to a play, adapted from the novel, by Mrs. Cecil Chesterton and Ralph Neale (1926). There and in other places Chesterton made clear that he intended Sunday to represent nature as it appears to a pantheist—seemingly indifferent to the struggle between good and evil. More recently Gary Wills, in his introduction to the 1975 edition of *The Man Who Was Thursday,* has analyzed the nightmare with shrewd insight.

Chesterton's only other fantasy novel, *The Ball and the Cross* (1909), is a more obvious Christian allegory. The story opens inside a "flying ship" piloted by Lucifer. His companion is an aged Bulgarian monk with whom he periodically argues. After Lucifer abandons the monk on the huge ball and cross of St. Paul's Cathedral in London, the monk manages to climb down to the top gallery, where he is arrested and taken to a mental hospital.

Below the cathedral, in the editorial office of a paper called *The Atheist,* its editor, James Turnbull, confronts Evan MacIan, a young Catholic who has been enraged by Turnbull's attack on the Virgin Mary. The two men vow to duel to the death, and most of the novel describes their wandering about in search of a spot where they can clash swords without being arrested. The ball on the cathedral is a symbol of Turnbull's rationalism, the cross a symbol of MacIan's faith. The duel, which the world does its best to prevent, is the eternal conflict between the two perspectives.

In the course of their sporadic fighting, Turnbull and MacIan grow fond of each other. After many adventures, melodramatic and romantic, the pair are tricked into entering the gardens of the asylum in which the old monk is imprisoned. The estate, a symbol of modern culture, is controlled by Lucifer.

In their dreams Turnbull and MacIan are taken separately in Lucifer's airship to London, where the devil offers each man a role in a political system that would seem to be to his liking. MacIan is shown an England ruled by a Christian dictatorship that sacrifices justice to obedience. Turnbull sees an England in the throes of a bloody secular revolution. Both men reject the hellish temptations by jumping out of the airship and waking up. Turnbull is now aware of the evil of needless political bloodshed, and MacIan recognizes the evil in what today would be called Christian fascism. (Unfortunately, in his later years, Chesterton was less perceptive than MacIan. Mussolini professed admiration for *The Man Who Was Thursday,* and Chesterton in return praised the Italian dictator in one of the most embarrassing of his books, *The Resurrection of Rome,* 1930.)

Lucifer's asylum, it turns out, contains all those who have seen MacIan and Turnbull fight. Leaders of the British government, under Lucifer's influence, have become so concerned over the extent to which the duel was stimulating interest in Christianity that they found it necessary to persuade the populace that the duel is a myth. The two antagonists—and everyone aware of their rash vows—have been declared insane.

After their dreams, Turnbull and MacIan realize the absurdity of their vows. MacIan sees his anger as excessive Christian zeal— the sort that inflamed the Inquisition. Turnbull sees his rage as a similar excess—the sort that produced the horrors of the French Revolution.

The asylum's inmates finally rebel and set fire to the buildings. When the aged monk emerges from his cell, the raging flames part like the Red Sea. The monk, a symbol of the church, forever old and young, walks down the path singing and laughing like a child. Lucifer escapes in his airship. When the fire subsides, MacIan sees his sword and Turnbull's in the ashes. They have fallen in the pattern of a cross.

Both fantasy novels are written in a style typical of Chesterton: with dazzling metaphors, rich alliteration, wordplay that resembles swordplay, and many statements that have the flavor of logical paradoxes. Almost all Chesterton's novels and stories convey metaphysical messages, with little effort to develop character—all his heroines talk and act alike—and *The Ball and the Cross* is no exception. It is seldom read today, partly because it is so explicit in its Catholic rhetoric, perhaps also because it contains touches of Chesterton's unconscious but irrepressible anti-Semitism.

Two of Chesterton's three plays are fantasies. *Magic,* written at the suggestion of his friend George Bernard Shaw, had a short run at London's Little Theatre in 1913 and a revival in London in 1942. It is about a nameless conjurer who has been hired by a simple-minded duke to entertain at his home. The duke, disturbed by his Irish niece's belief in fairies, hopes that seeing good prestidigitation will persuade her that real magic does not exist.

The niece's American brother, a confirmed materialist, angers the conjurer by explaining how his apparatus works. Soon mysterious events begin to occur. A picture on the wall sways. A chair overturns. The brother thinks this is part of the magic act. When a red lamp seen through a window turns blue, the brother rushes into the rainy night to learn how the trick was done. The frustration of being unable to find out triggers a nervous collapse. To restore the brother's sanity, his sister begs the conjurer to explain how he did the trick. But he cannot explain. The magic was genuine. At one time he had dabbled in the occult, and in his anger at the brother he had called on evil spirits. To help the brother regain his sanity, the conjurer fabricates a natural explanation.

Chesterton's other fantasy play, *The Surprise,* was written in 1930 and published in 1952. In the play an "author" owns a troop of robot actors who speak exactly as programmed. They perform a play within a play that ends happily, all its characters having behaved with the highest morality. Then, as a result of the prayer of a wandering Franciscan monk (who happens to see the play), the robots become real people with free will. When the play is repeated, their motives turn base and the play takes an ugly turn. *The Surprise* ends when the author, his head bursting through a top portion of the scenery, shouts: "And in the devil's name, what do you think you are doing with my play? Drop it! Stop! I am coming down." The play is an obvious parable of the Incarnation.

There are eleven books of short stories by Chesterton, most of them mysteries, though only the five books about his crime-solving priest, Father Brown, continue to be widely read. A few fantasies that he wrote when very young are in a posthumous collection, *The Coloured Lands* (1938). Only the title story is memorable. It opens with a little boy named Tommy sitting on the lawn of his house, bored and lonely. Even the landscape's colors seem dull. Suddenly a young man wearing blue spectacles leaps over a hedge and hands Tommy his glasses. The boy is entranced by the azure world he sees and even more delighted when he looks through spectacles of other colors.

The stranger tells Tommy that as a boy he, too, grew tired of life. A wizard granted his desire for a different world by transporting him to a place where everything was blue. When he became tired of blue, the wizard made everything green, and when he got bored with green, everything turned yellow, and finally red. (Chesterton was anticipating the colored lands of L. Frank Baum's Oz.) Unfortunately, "in a rose-red city you cannot really see any roses." The stranger tells Tommy that soon he grew tired of red. "Well," the wizard told the stranger, "you don't seem very easy to please. If you can't put up with any of these countries, or any of these colors, you shall jolly well make a country of your own."

The stranger goes on to say that he soon found himself, with a large supply of paints, in front of an enormous blank space. After splashing the top of the space with blue, he put a square of white in the middle, and spilled some green along the bottom. Having learned that the secret of red is to have a small amount of it, he dabbed a few spots of red above the green. Slowly he realized what he had done. He had created the very landscape at which he and Tommy were gazing. Having finished his tale, the stranger hops back over the hedge, leaving Tommy staring at the scenery "with a new look in his eyes."

The new look is, of course, Chesterton's lifelong mystical vision. Sane and happy people, Chesterton believed, should look at the world with emotions of wonder, surprise, and gratitude. They should find it more astonishing than any magic show, more colorful than Oz, more fantastic than any fantasy, surrounded on all sides and suffused throughout with awesome, unthinkable mystery.

Selected Bibliography

Fantasy Works of G. K. Chesterton

The Man Who Was Thursday: A Nightmare. Bristol, England: Arrowsmith, 1908. (Advance issue 1907) New York: Dodd, Mead, 1908. (Novel)

The Ball and the Cross. New York: John Lane, 1909. London: Wells Gardner, Darton, 1910. (Novel)

Magic: A Fantastic Comedy. London: Secker, 1913. New York: Putnam, 1913. (Play)

The Man Who Was Thursday. Adapted by Mrs. Cecil Chesterton and Ralph Neale. London: Ernest Benn, 1926. (Play)

The Coloured Lands. London and New York: Sheed and Ward, 1938. (Stories and drawings)

The Surprise. Preface by Dorothy L. Sayers. London and New York: Sheed and Ward, 1952. (Play)

Critical and Biographical Studies

The Chesterton Society, devoted to the study and promotion of interest in Chesterton and his writings, was founded in England in 1974. Since then it has published *The Chesterton Review* (St. Thomas More College, Saskatoon, Saskatchewan, Canada), a periodical edited by Ian Boyd.

Chesterton, G. K. *Autobiography.* London: Hutchinson, 1936. As *The Autobiography* of *G. K. Chesterton.* New York: Sheed and Ward, 1936.

Dale, Alzina Stone. *The Outline of Sanity: A Life of G. K. Chesterton.* Grand Rapids, Mich.: Eerdmans, 1982.

Sullivan, John J., ed. *G. K. Chesterton: A Centenary Appraisal.* New York: Harper and Row, 1974.

Ward, Maisie. *Gilbert Keith Chesterton.* New York: Sheed and Ward, 1943.

———. *Return to Chesterton.* New York: Sheed and Ward, 1952.

Wills, Gary. *Chesterton: Man and Mask.* New York: Sheed and Ward, 1961.

———. Introduction to *The Man Who Was Thursday.* New York: Sheed and Ward, 1975.

Bibliographical Studies

Sullivan, John. *G. K. Chesterton: A Bibliography.* London: University of London Press, 1958.

————. *Chesterton Continued: A Bibliographic Supplement.* London: University of London Press, 1968.

————. *Chesterton Three: A Bibliographical Postscript.* Bedford, England: Vintage Publications, 1980.

15 The Fantasies of Lord Dunsany

Lord Dunsany is generally considered Great Britain's finest twentieth-century writer of fantasy. Most of his novels, short stories, and plays are fantasies, written in a remarkable style that is rich in the musical cadences of the King James Bible and that swarms with exotic names for gods, kings, and lesser mortals, as well as for places and things. The names usually came effortlessly to Dunsany; he wrote in his autobiography that when they required conscious thought, "the name has always been uninteresting, unconvincing, and as though it were not the real name."

Edward John Moreton Drax Plunkett, Lord Dunsany, was born in London on July 24, 1878; his ancestors were Irish, having long occupied Dunsany Castle, in County Meath, about twenty miles from Dublin. After an education at Eton and Sandhurst, he became the eighteenth Baron Dunsany in 1899 when his father died. During the Boer War he fought in the Coldstream Guards, and in World War I he was a captain in the Fifth Royal Inniskilling Fusiliers. Hunting and chess were his most passionate avocations. He was a master at chess, at one time champion of Ireland, and the creator of many whimsical chess problems. He also was active in the early days of the Abbey Theatre in Dublin, where his name was frequently linked with those of William Butler Yeats, John Millington Synge, and Lady Gregory. It was in Dublin that he died on October 25, 1957, although he and his wife had long lived in Shoreham, Kent, after giving Dunsany Castle to their only child, Randal.

Dunsany's first book, *The Gods of Pegāna* (1905), introduces an elaborate mythology, rivaled in color and subtlety only by the mythologies of J. R. R. Tolkien and James Branch Cabell. *Time and*

This article originally appeared in *Supernatural Fiction Writers: Fantasy and Horror,* vol. 1, edited by E. F. Bleiler, and is reprinted here with the publisher's permission. © 1985 by Charles Scribner's Sons.

the Gods (1906) contains further accounts of what Dunsany, in a one-sentence preface, calls "the things that befell gods and men in Yarnith, Averon, and Zarkandhu, and in the other countries of my dreams."

The *Sword of Welleran* (1908) and *A Dreamer's Tales* (1910) move away from Pegāna to new realms of fancy. *The Book of Wonder* (1912) is unique among Dunsany's fictional works in the way it was written. Instead of giving his friend Sidney Herbert Sime stories to illustrate, as was his usual practice, Dunsany wrote stories to fit some drawings by Sime. Subtitled "A Chronicle of Little Adventures at the Edge of the World," his wonder tales are introduced with another one-sentence preface: "Come with me, ladies and gentlemen who are in any wise weary of London: come with me: and those that tire at all of the world we know: for we have new worlds here."

Stories in *Tales of Wonder* (1916) are richer in humor than any of Dunsany's other works of supernatural fiction, although all of his fiction was guided, as Dunsany put it in his autobiography, "by two lights that do not seem very often to shine together, poetry and humor." The book includes such famous tales as "The Exiles' Club," in which exiled kings are merely the waiters who serve the abandoned Greek gods who live above, and "The Three Sailors' Gambit," which tells how three seafaring men, with the help of a magic crystal ball, become unbeatable at chess.

Tales of Three Hemispheres (1919) and *The Man Who Ate the Phoenix* (1949) are two other collections of fantasy stories. Dunsany also wrote five books of stories about the adventures of Jorkens: *The Travel Tales of Mr. Joseph Jorkens* (1931), *Mr. Jorkens Remembers Africa* (1934), *Jorkens Has a Large Whiskey* (1940), *The Fourth Book of Jorkens* (1948), and *Jorkens Borrows Another Whiskey* (1954). Many of these tales are fantasies, but because most are told at the Billiards Club, in London, by a man who is a notorious liar, perhaps they should be classified more as humorous tall tales than as pure fantasy. *The Little Tales of Smethers* (1952) are not fantasies but mystery stories. The few told by Smethers include the much-anthologized tale of cannibalism "Two Bottles of Relish." "The New Master," about a jealous chess-playing robot that poisons its inventor, can be labeled science fiction.

Dunsany's first novel, *The Chronicles of Rodriguez* (1922), is a picaresque tale of chivalry, romance, and adventure set in the golden age of Spain. Young Rodriguez goes in search of a just war, with hopes of obtaining a wife and a castle. In an emerald-encrusted scabbard he carries a sword and, on his back, a mandolin that he plays as skillfully as he wields the blade. Like Don Quixote, he travels with a faithful, simpleminded servant.

After many fantastic adventures, Rodriguez finds the war he seeks but fails to obtain either wife or castle. A man he had earlier saved from hanging turns out to be the leader of a band of merry, green-clad archers who control the forest of Shadow Valley in the manner of Robin Hood and his men. In gratitude, the green bowmen build for Rodriguez a great castle. Serafina, with whom he has fallen in love, accepts his proposal, and (in Dunsany's words) they "lived happily ever after."

The novel is beautifully written, filled with superb descriptions of nature, shrewd philosophical asides, and sardonic humor. Pure fantasy enters in chapter 3, when Rodriguez and his servant visit the House of Wonder, where an evil magician lives. Through one of his two magic windows, they see Spanish wars of the past not as history books relate them but as they actually were. Through the other window they see battles of the future. No longer do men fight hand to hand, with small loss of life. Because of the ingenuity of science, thousands of innocents are mangled and killed by gunpowder and the new machines of mass destruction. (The device of a magic window through which one can see distant times and places had been used by Dunsany earlier in a short story, "The Wonderful Window," and would be used again decades later in a radio play, *Golden Dragon City*.) Before Rodriguez and his servant leave, the magician sends them on an out-of-body journey past Venus and Mercury to the interior of the sun.

The setting of Dunsany's second novel, *The King of Elfland's Daughter* (1924), is the country of Erl. Not far away, separated from Erl by a region of perpetual twilight, are the glowing fields of Elfland. It is a region of supernal beauty, surrounded by pale-blue mountains, where time passes slowly and no person ages or dies. The king of Elfland, who sits on a throne of mist and ice, had once been married to a mortal. Their only child, Lirazel, is now Elfland's lovely princess.

Young Alveric of Erl, carrying a sword enchanted by a friendly witch, journeys to Elfland to win the hand of Lirazel. He succeeds in this quest, taking the princess to Erl, where she bears their son, Orion. Soon Lirazel, finding it difficult to understand the quaint customs of mortals, is longing for her father and her homeland. Eager to have his daughter back, the king of Elfland sends to her a troll who carries a rune against which Alveric's enchanted sword is powerless. The rune magically transports Lirazel home. When the heartbroken Alveric goes in search of her, the king withdraws Elfland to a more distant region.

The years pass quickly. Orion grows to become a mighty hunter, and several chapters tell how he tracks and kills a white unicorn that had strayed from Elfland. At times he can hear, in Lord Tennyson's familiar lines, "the horns of Elfland faintly blowing."

In Elfland the half-mortal Lirazel grieves for her husband and son and for the strange customs of Erl that she had come to love. Seeing her unhappiness at being caught between two worlds, the king reunites her with Alveric and Orion by extending the borders of Elfland. Like "unearthly foam," the enchanted land pours over all the fields of Erl except for a tiny region surrounding a "Christom" friar, for whom the magic ways of Elfland are evil.

In "Idle Days on the Yann," a story in *A Dreamer's Tales,* Dunsany introduces a phrase that he loved to repeat: "beyond the fields we know." In *The King of Elfland's Daughter* this phrase and the phrase "only told of in song" recur like musical refrains. Dunsany considered this novel closer to poetry than any of his other works of fiction.

The Charwoman's Shadow (1926), Dunsany's third novel, returns to Spain's golden age. Hard times have come to a noble family in Shadow Valley. Hoping to obtain money for his daughter's dowry, the father sends his son, Ramon Alonzo, to an ageless magician who owes the family a favor. There Ramon is to serve as an apprentice, studying the black arts, especially the alchemical art of making gold from lead.

The magician is pleased to instruct Ramon, but for this he extracts a curious price. Ramon must give him his shadow. The magician collects human shadows, keeping them in a box that can be opened only by chanting three secret Chinese syllables. The magician often sends his shadows on long trips into outer space, where they join evil spirits to engage in dreadful deeds known only to himself. Living in the magician's strange house is an aged, wrinkled charwoman. Many decades ago, when she was a poor girl in a nearby village, the magician had bought her from her parents to be his only servant. In return for giving her the inability to die, he had taken her shadow. Now she is old and miserable, longing for her shadow and hopelessly trapped by the magician's evil magic.

Ramon and the old crone become inexplicably fond of one another, and Ramon vows that he will restore her shadow to her. Foolishly he allows the magician to take his own shadow when he is promised a false one exactly like the real one. Later he discovers to his horror that the false shadow never alters in length, even when the sun is low.

By subtle stratagems, aided by the combinatorial art of the thirteenth-century Spanish mystic Ramon Lull, Ramon Alonzo manages to learn the three secret syllables, "Ting Yung Han," that open the keyless padlock on the box of shadows. He finds his own shadow—it quickly rejoins his feet—but not the charwoman's. However, among the writhing gray forms is one of a slender young girl so beautiful that Ramon instantly falls in love with its shape. It turns out to be the charwoman's shadow. Ramon takes it to her, and the pair escape. In a sunlit field, as Ramon attaches the shadow to her feet, he turns his head, unable to bear the pain of seeing the shadow acquire the shape of a hag. To his amazement, the opposite occurs. The shadow, more real than the body's substance, transforms the withered charwoman into a laughing girl of seventeen.

Ramon takes her home with him. The king of Shadow Valley pardons her low birth, and (again in Dunsany's words) "she and Ramon Alonzo lived happily ever after." The magician proves to be none other than Pan. He leaves Spain and, after wandering for a time over Europe, abandons the earth entirely.

The goat-footed god is back again in Dunsany's fourth novel, *The Blessing of Pan* (1927). Elderick Anwrel, the plump vicar of the English village of Wolding, Kent, is disturbed that his parishioners are being influenced by unearthly, flutelike music coming from Wold Hill. Wild and unfamiliar, the tune agitates the mind and stimulates strange desires.

A local farm boy, Tommy Duffin, plays the tune on pipes that he cut from reeds. Anwrel's inquiries disclose that Tommy's parents had been married by the vicar's mysterious predecessor, the Reverend Arthur Davidson. After the ceremony, Davidson had cast a spell on the newlyweds by speaking to them in a curious language that they could not understand. Later, on a moonlight night, when Davidson was seen dancing without his boots in the vicarage garden, it was observed that he had an extra joint on each leg between ankle and knee. The next day, Davidson abruptly left the village. As Anwrel correctly surmises, Davidson is Pan.

The music from Tommy's pipes draws the young people of Wolding to Wold Hill, where twelve large black stones surround a flat stone used as an altar for pagan sacrifices in the days before Saint Ethelbruda. Not far away is the saint's tomb, where villagers used to go for miraculous healing of their warts. Now the tomb has lost this power. Residents of Wolding have grown careless of their gardens and houses. Lawns have become invaded by wildflowers, moss, and rabbits. Even the foxes venture down the slopes to the village. A schoolteacher stops teaching arithmetic.

Anwrel's efforts to halt the spread of paganism are of no avail. While he preaches a stirring sermon on Christian faith, the pipes of Tommy Duffin are heard. One by one his parishioners leave the service, first the young girls and men, then the elders, until only the vicar's wife remains. Then she, too, leaves.

On Wold Hill the villagers plan to sacrifice a bull. Late that night, moved by an uncontrollable impulse, Anwrel fashions a crude ax and heads for the Old Stones. A flame is burning on the flat rock, and the shape of Pan can be seen dimly in the woods above. At dawn, four young men bring a bull to the altar: "There slept along Anwrel's arms, and were not yet withered, muscles with which he had rowed when thirty years younger." He swings the ax, and the bull is slain.

Great Pan once died, but now it is Christianity that dies in Wolding. The pagan rituals take over, with Tommy as the local druid priest. Anwrel is not unfrocked, but he leaves the vicarage to live in a hut nearer Wold Hill. Machinery vanishes in the town. Even money disappears, as the villagers revert to barter and the old crafts:

> To this queer community recruits came rarely from the lands beyond Wold Hill, from the world and the ways we know; rarely, but yet they came. For those pipes of Tommy Duffin playing often in summer evenings would drift their music perhaps a mile on still air, perhaps much further, till the notes would come to some hill beyond Wolding's woods, where a picnic party from London would be sitting on a Sunday afternoon, throwing broken bottles for fun in the mint and thyme. It had to be a still evening; and even then not a sound would come so far but to ears that, weary with the same old mumble that some machine told over and over and over, were listening for something utterly strange and new. To such ears, as they leaned towards it, that music might faintly reach from where Tommy Duffin played on the slopes of Wolding. After that some girl would slip away alone from the lemonade and gramophone, and was seldom found till long after; and if they ever found her at all she would no longer seem to understand cities. And tired shopwalkers, sick of salesmanship, would sometimes find their way there, pushing through saplings and briar on a Bank-holiday, never to leave the valley for London any more. (Chap. 35)

The death of modern cities and the slow return of nature, symbolized by Pan, were some of Dunsany's favorite themes. In "The

Tomb of Pan," the last short story of his *Fifty-One Tales* (1915), Pan dies and is buried: "But at evening as he stole out of the forest, and slipped like a shadow softly along the hills, Pan saw the tomb and laughed." *The Blessing of Pan* is vintage Dunsany, expressing in musical language his love of uncontaminated nature, his hatred of modern business and technology, his contempt for Christianity, and his fondness for forgotten gods. "I am sad, master, when the old gods go," remarks a servant in Dunsany's play *If* (1922). "But they are bad gods, Daoud," says his master. Daoud replies, "I am sad when the bad gods go."

Dunsany also wrote about the Ireland he knew, and *The Curse of the Wise Woman* (1933) is his most autobiographical novel. An elderly narrator recalls his happy boyhood in Meath, when he roamed the marshes hunting and fishing while Ireland was shaken by political conflict. An old witch pronounces a curse on the Peat Development Syndicate, which tries to have the marshes drained so that it can compress the soil and sell it as coal. It is the working out of this curse that gives the novel its fantasy element.

The men who work for the syndicate regard Mrs. Marlin, the witch, as crazy but harmless. A great storm hits the region. Throughout an entire night, buffeted by violent winds and rain, Mrs. Marlin shouts her curses in a tongue that no one understands. Next morning the sky is blue and there is an ominous calm. "The bog is coming," says Mrs. Marlin just before she dies. And come it does, pouring over the area and burying the peat factory under eight feet of water. Fifty years later, it is still submerged, except for some roof ornaments on which birds perch.

Two later books—actually collections of related short stories, although printed in the form of novels—concern the transfer of human minds into animals'. *My Talks with Dean Spanley* (1936) is a small book about a man who, under the influence of wine, recalls episodes in his previous incarnation as the loyal watchdog of an English estate. *The Strange Journeys of Colonel Polders* (1950) centers on Pundit Sinadryana from Benares, who belongs to a men's club in Chelsea. Using incense and magic spells, he has the power to send Colonel Polders' spirit into any animal he chooses. The colonel entertains his fellow club members with tales of his adventures as a fish, bird, dog, pig, fox, tiger, sheep, butterfly, moth, cat, antelope, mouse, monkey, camel, snail, flea, and other forms of life. His listeners are skeptical until the Indian sorcerer turns one into a frog, another into a coolie, and a third into a squirrel.

Rory and Bran (1936) is not fantasy. Although Bran is a dog, Dunsany never explicitly says this, and unastute readers are capable of finishing the novel without realizing that Bran is not human. *Mr. Faithful* (1935), a play that also belongs to Dunsany's "canine period," concerns a man who takes a job as a dog.

Dunsany wrote one science-fiction novel, *The Last Revolution* (1951). It tells how computers, after acquiring intelligence, free will, and the ability to make copies of themselves, threaten to take over the civilized world. In one memorable scene the narrator, playing chess with a crablike robot, suddenly realizes that he is playing not against a program but against something that is alive and possesses a mind far superior to his own.

A number of collections of Dunsany's short stories have been published posthumously, but only *Ghost of the Heaviside Layer and Other Fantasms,* edited by Darrell Schweitzer (1980), contains writings not previously in any book. Among them are both a Jorkens story based on a chess problem with an unorthodox solution and one of Dunsany's many stories about pacts with the devil. In "Told Under Oath" a man spins a fantastic yarn about how Satan has given him the power of never hitting a golf ball without making a hole in one. Instead of enjoying this ability, he is prevented from playing golf with anyone. The story ends with a delightful twist on the old liar paradox of logic. When asked what price he paid for the gift, he replies that Satan "extorted from me my power of ever speaking the truth again."

Dunsany's first success as a playwright came in 1909 when his one-act play *The Glittering Gate,* written at the request of Yeats, was produced at the Abbey Theatre. Two recently deceased burglars, Bill and Jim, find themselves locked outside the gates of heaven. They manage to force the lock, but when the enormous gate swings open, nothing is there except an empty blue void filled with "bloomin' great stars."

"There *ain't* no heaven, Jim," says Bill, while cruel laughter howls louder and louder offstage as the curtain falls. This short play was the beginning of Dunsany's distinguished career as an Irish dramatist. So popular were his fantasy plays in both England and the United States that during the first half of his life Dunsany was more highly regarded as a dramatist than as a writer of fiction or poetry. In 1916 five of his plays ran concurrently on Broadway.

The Glittering Gate is in Dunsany's first book of dramas, *Five Plays* (1914). The book also contains his highly praised play of the supernatural, *The Gods of the Mountain,* with its startling line "Rock should not walk in the evening." Later books of dramatic works are *Plays of Gods and Men* (1917), *Plays of Near and Far*

(1922), *Alexander and Three Small Plays* (1925), *Seven Modern Comedies* (1928), *The Old Folk of the Centuries* (1930), *Lord Adrian* (1933), *Mr. Faithful* (1935), and *Plays for Earth and Air* (1937). "Air" in the last title refers to dramas written for British radio.

Although *If*, Dunsany's most successful full-length play, is based on the idea of diverging time paths, it is more properly fantasy than science fiction. A happily married man living in a London suburb misses his morning train. Ten years later he obtains a magic crystal that allows him to go back in time. He returns to the morning on which he missed the 8:15, but now he catches it. A meeting on the train with a beautiful young woman leads to ten years of adventures in the Far East that end disastrously. Not until the crystal is shattered is he able to return to his former time path.

Dunsany hoped to repeat the success of *If* with his full-length play *Lord Adrian*, but it was not well received. More science fiction than fantasy, it concerns an aged duke who wants to marry Bessie, his young secretary. A "monkey gland" graft—a rejuvenation process that was making a medical stir at the time (Yeats actually underwent such an operation)—restores his youth and virility. Bessie marries him and they have a son, Adrian.

At age twenty, handsome and charming, Lord Adrian distresses his parents by behaving and talking in increasingly odd ways. He likes to walk barefooted through the woods. He talks to animals. He refuses to attend church. One Sunday, while church bells ring, he places flowers at the foot of a statue of Pan. He shocks his father by expressing hatred for the sport of hunting and his belief that humanity is destroying animals who are in many ways superior to men. He vows to teach the wild beasts the secret of fire so that they can fight back.

The estate's gamekeeper owns a once-faithful dog that he kills when it stops obeying him. One day in the woods the gamekeeper finds Lord Adrian lighting little fires with flints and explaining to foxes and badgers how they can do the same. Adrian leaps into a tree, swings up like Tarzan, rips off a huge branch, and drops to the ground. To protect himself, the gamekeeper fires his gun. It was no human soul he killed, he reports to the duke. But now, as the curtain descends, the duke is too senile to comprehend the tragedy. Unlike in *The Blessing of Pan*, nature here has lost a round in its eternal battle against corrupting civilization.

One of Dunsany's poems, "Waiting," speculates on which will survive—London, "with its awful cluster of towns," or the wild woods, over which the borders of London are steadily creeping. A "grey and reverend oak" gives the answer:

> "Yes, we have heard of it:
> We have known such cities of old:
> We stand and we dream a bit,
> And its weald again and the wold."

As in all good fantasy, it is easy to find allegories in Dunsany's plays and fiction, but he steadfastly denied that he ever put them there intentionally. "Cleverness later led some people to look for allegories in my plays," he writes in his autobiography, "and once you start looking for allegories you are lost in a maze that has no center." In a letter printed in *Dunsany the Dramatist* (1917), by E. H. Bierstadt, Dunsany says: "Don't let them hunt for allegories. I may have written an allegory at some time, but if I have, it was a quite obvious one, and as a general rule, I have nothing to do with allegories. . . . When I write of Babylon, there are people who cannot see that I write of it for love of Babylon's ways, and they think I am thinking of London still and our beastly Parliament. Only I get farther east than Babylon, even to Kingdoms that seem to lie in the twilight beyond the East of the World."

Nine books of Dunsany's poetry were published during his lifetime, and hundreds of his poems remain buried in periodicals and unpublished papers. Dunsany disliked all modern verse for its obscurity and lack of melody; he especially detested the work of Ezra Pound and T. S. Eliot. Poetry, he once said in a speech, should ring like bells; modern verse does nothing but "just klunk." The living poet he most admired was Walter de la Mare. Dunsany's most notable lecture on poetry was published in 1918 as a small book titled *Nowadays*.

Dunsany's nonfiction includes a book of World War I sketches, *Unhappy Far-Off Things* (1919); a three-volume autobiography, *Patches of Sunlight* (1938), *While the Sirens Slept* (1944), and *The Sirens Wake* (1945); *My Ireland* (1937); several introductions to books by others; and numerous articles in popular and obscure periodicals.

No one has better described the magic quality of Dunsany's visions than Yeats in his preface to *Selections from the Writings of Lord Dunsany* (1912):

> Yet say what I will, so strange is the pleasure that they give, so hard to analyse and describe, I do not know why these stories and

plays delight me. Now they set me thinking of some old Irish jewel work, now of a sword covered with Indian Arabesques that hangs in a friend's hall, now of St. Marks at Venice, now of cloud palaces at the sundown; but more often still of a strange country or state of the soul that once for a few weeks I entered in deep sleep and after lost and have ever mourned and desired.

Yeats's comments apply equally well to the art of Dunsany's illustrator, Sidney Herbert Sime, whose fame as a fantasy artist at one time rivaled Dunsany's fame as a writer of fantasy. Sime illustrated many of Dunsany's early books of short stories and provided frontispieces for four of his novels.

Although Sime was greatly admired in his early years by many art critics, he died despondent and penniless, his reputation in total eclipse. Many of his original pictures hang in Dunsany Castle, others are in the Sime Memorial Gallery at Worplesdon, Surrey, where he and his wife lived. Today his reputation, like that of Dunsany, is enjoying a strong revival among fantasy enthusiasts, but both men have been almost forgotten by the general public and continue to be ignored by leading critics of the literary and art worlds.

Selected Bibliography

Fiction and Plays of Lord Dunsany

The Gods of Pegāna. London: Elkin Mathews, 1905. Boston: Luce, 1916. (Short stories)

Time and the Gods. London: Heinemann, 1906. Boston: Luce, 1913. (Short stories)

The Sword of Welleran and Other Stories. London: George Allen, 1908. Boston: Luce, 1916. (Short stories)

A Dreamer's Tales. London: George Allen, 1910. Boston: Luce, 1916.

The Book of Wonder. A Chronicle of Little Adventures at the Edge of the World. London: Heinemann, 1912. Boston: Luce, 1913. (Short stories)

Five Plays. London: Grant Richards, 1914. New York: Mitchell Kennerly, 1914.

Fifty-One Tales. London: Elkin Mathews, 1915. New York: Mitchell Kennerly, 1915. (Short-short stories)

Tales of Wonder. London: Elkin Mathews, 1916. *As The Last Book of Wonder.* Boston: Luce, 1916.

Plays of Gods and Men. Dublin: Talbot Press, 1917. Boston: Luce, 1917.

Tales of Three Hemispheres. Boston: Luce, 1919. London: Unwin, 1920.

The Chronicles of Rodriguez. London: Putnam, 1922. As Don Rodriguez, *Chronicles of Shadow Valley.* New York: Putnam, 1922.

If: A Play in Four Acts. London and New York: Putnam, 1922.

Plays of Near and Far. London: Putnam, 1922. New York: Putnam, 1923.

The King of Elfland's Daughter. London and New York: Putnam, 1924.

Alexander and Three Small Plays. London: Putnam, 1925. New York: Putnam, 1926.

The Charwoman's Shadow. London and New York: Putnam, 1926.

The Blessing of Pan. London: Putnam, 1927. New York: Putnam, 1928.

Seven Modern Comedies. London: Putnam, 1928. New York: Putnam, 1929.

The Old Folk of the Centuries. London: Elkin Mathews and Marrot, 1930. (Play)

The Curse of the Wise Woman. London: Heinemann, 1933. New York: Longmans, Green, 1933.

Lord Adrian: A Play in Three Acts. Waltham Saint Lawrence, England: Golden Cockerel Press, 1933.

My Talks with Dean Spanley. London: Heinemann, 1936. New York: Putnam, 1936. (Short novel)

Plays for Earth and Air. London: Heinemann, 1937.

The Man Who Ate the Phoenix. London: Jarrolds, 1949. (Short stories)

The Strange Journeys of Colonel Polders. London: Jarrolds, 1950.

Modern reprint editions

At the Edge of the World. Edited by Lin Carter. New York: Ballantine, 1970. (Short stories)

Beyond the Fields We Know. Edited by Lin Carter. New York: Ballantine, 1972. (Short stories)

Gods, Men and Ghosts: The Best Supernatural Fiction of Lord Dunsany. Edited by E. F. Bleiler. New York: Dover, 1972. (Short stories)

Over the Hills and Far Away. Edited by Lin Carter. New York: Ballantine, 1974. (Short stories)

Ghosts of the Heaviside Layer and Other Fantasms. Edited by Darrell Schweitzer. Philadelphia: Owlswick Press, 1980. (Short stories and *Lord Adrian*)

Critical and Biographical Studies

Amory, Mark. *Biography of Lord Dunsany.* London: Collins, 1972.

Bierstadt, Edward Hale. *Dunsany the Dramatist.* Boston: Little, Brown, 1917.

Boyd, Ernest. "Lord Dunsany: Fantaisiste." *In Appreciations and Depreciations.* New York: Lane, 1918.

Colum, Padraic. Introduction to *A Dreamer's Tales.* New York: Modern Library, 1917.

de Camp, L. Sprague. "Two Men in One: Lord Dunsany." *In Literary Swordsmen and Sorcerers: The Makers of Heroic Fantasy.* Sauk City, Wis.: Arkham House, 1976.

Dunsany, Lord. *Patches of Sunlight.* London: Heinemann, 1938. (Autobiography)

———. *While the Sirens Slept.* London: Jarrolds, 1944. (Autobiography)

———. *The Sirens Wake.* London: Jarrolds, 1945. (Autobiography)

Gardner, Martin. "Sidney Sime of Worplesdon." *Arkham Sampler,* Autumn 1949. (Reprinted in Gardner, *Order and Surprise.* Buffalo: Prometheus Books, 1983.)

———. Introduction to *A Dreamer's Tales.* Philadelphia: Owlswick Press, 1979.

Harris, Frank. "Lord Dunsany and Sidney Sime." *In Contemporary Portraits, Second Series.* New York: 1919. (Privately printed)

Littlefield, Hazel. *Lord Dunsany: King of Dreams.* New York: Exposition Press, 1959.

Yeats, William Butler. Preface to *Selections from the Writings of Lord Dunsany.* Churchtown, Ireland: Cuala Press, 1912.

Playing with Mathematics

For decades some of us have been struggling to persuade mathematics educators that the best way to motivate youngsters is to give them tasks they actually enjoy. By first capturing a class's interest with a good puzzle, paradox, game, model, toy, joke, or magic trick, the teacher can lead students into significant ideas without the students even knowing they are learning.

I could give a thousand examples, but there is space for only two that do not require pictures. Suppose you are teaching multiplication. Instead of offering dull word problems, such as how much do ten dozen 17-cent stamps cost, try this approach. Write 12345679 (note the missing 8) on the blackboard, then ask a student to name any digit from 1 through 9. Suppose it is 3. Tell the class that if they multiply 12345679 by 27, the answer will be a big surprise. Children love surprises. You can see the astonishment on their faces when they get the answer: 333,333,333. Repeat the stunt with another digit, say 8. Now the multiplier is 72 and the product is 888,888,888.

How do you know what multiplier to give? The class is intrigued, so you give away the secret. Dividing 111,111,111 by 9 yields 12,345,679, so of course multiplying that number by 9 will restore 111,111,111. Obviously, if 111,111,111 is multiplied by any digit K, the product will be KKKKKKKKK. To select the proper multiplier, simply take the product of the chosen digit and 9. You have just introduced the dreaded "associative law" of the new math: $(ab)c = a(bc)$. The class is fascinated because it is a trick they can show on a pocket calculator to their parents and friends. The sudden appearance of nine identical digits in the readout is unfailingly amazing.

Here is another simple example. Write on the blackboard, the

This article originally appeared in *The Washington Post,* 5 April 1987, and is reprinted here with permission. © 1987 by The Washington Post.

lucky dice numbers 7 and 11, and the unlucky 13. Ask a child to write any three-digit number on the blackboard, then repeat the digits to make a six-digit number of the form ABCABC. Assume ABC = 382.

Ask the class to divide 382,382 by 7. You predict that there will be no remainder. Sure enough, the quotient comes out 54,626. Ask them to divide this by 11. Again, no remainder. The quotient is 4,966. The trick is getting spookier. Now a final division by 13. For the third time there is no remainder, but now there is a bigger surprise. The result is 382, the original number!

Does the trick work with any number of the form ABCABC? You can be sure the class would like to know. Give them some time to figure it out for themselves, and if they cannot you explain. ABC times 1,001 must of course produce ABCABC. Now the prime divisors of 1,001 are 7, 11, and 13, so 7 × 11 × 13 = 1,001. If (7 × 11 × 13) × ABC = ABCABC, it is no mystery that when ABCABC is divided by 7, 11 and 13, the quotient will be ABC. You have underscored the fact that division is the inverse of multiplication.

Raymond Smullyan, a world expert on logic and set theory, once taught a course in elementary geometry. He asked a student to draw an arbitrary right triangle on the blackboard. Smullyan drew a square on all three sides, then posed the following question. If the three squares were sheets of pure gold, would the large square be worth more or less than the sum of the two smaller ones? Some students guessed more, some less. The class was astounded when Smullyan assured them that the worth would always be the same regardless of the triangle's proportions. Now they were interested enough to stay awake while he gave a simple proof of the famous Pythagorean theorem.

There is such a rich literature on this approach to teaching that the National Council of Teachers of Mathematics sells a four-volume *Bibliography of Recreational Mathematics,* by William Schaaf, and the magazine *Mathematics Teacher* has been steadily increasing the amount of such material on its pages. Ten years ago it was considered "enrichment" material to be given only sparingly to brighter students, but now teachers are beginning to see (I hope) that it should pervade all their instruction if they want to keep students from daydreaming.

Unfortunately, there is a reason why the movement of recreational math into classrooms has been so glacial. It is useful only to teachers who themselves are smitten by the wonder and beauty of mathematical patterns and who find teaching them an exciting, satisfying experience.

PART TWO
REVIEWS

Polywater

In 1962 an obscure Soviet chemist claimed to have found a strange new kind of water. No one paid much attention to him until 1968, when Boris Deryagin, a top Soviet chemist, announced that careful testing had validated the earlier claim. Polywater, as it was soon called, seemed easy to make. One merely allowed distilled water vapor to condense in hairlike capillary tubes. For unknown reasons the liquid apparently acquired a polymeric molecular structure (hence its name) with fantastic properties. It became superdense and jellylike. It froze and boiled at odd temperatures.

Because the new water had both theoretical and practical applications that were revolutionary, British and U.S. chemists were understandably excited. Deryagin was, after all, too respected to be dismissed as a crank. Polywater, declared J. D. Bernal of London's Birkbeck College, was the "most important physical-chemical discovery of this century."

"Creepy water" (as the *Washington Post* called it) was believed to have important military uses. Fearful of a polywater gap, the Army, the Navy, and other U.S. agencies started dishing out grants, and the bandwagon began to roll. The bizarre story is skillfully documented by Felix Franks, an English biochemist, in *Polywater* (MIT Press, 1981).

From the start, establishment chemists were skeptical. Unable to produce the miracle water, they maintained it was just ordinary H_2O contaminated by impurities. Franks thinks some of them overdid their sarcasm. (A Purdue scientist called the water "polycrap.") But several hundred honorable chemists rushed into print with reports based on careless experiments and with all sorts of unguarded

This review originally appeared in *Science Digest,* September 1981, and is reprinted here with permission.

statements. One scientist warned that polywater, like the "ice-nine" in Kurt Vonnegut's novel *Cat's Cradle,* might change all water to polywater and make our planet as uninhabitable as Venus!

The epidemic crested in 1970, then slowly ebbed as evidence for contamination grew. Ordinary human perspiration turned out to be one contaminant, but the main offender was silicon "leached" from the glass or quartz tubes in which polywater supposedly formed. The end came in 1978 when Deryagin himself conceded that for a decade he had been studying nothing more than dirty water.

How did it happen that millions of dollars were squandered on this quixotic quest? Franks faults government agencies for premature funding and technical journals for overpermissiveness. He faults experimenters for self-deception and theoreticians for forgetting Sherlock Holmes's advice: "It is a capital mistake to theorize before one has data." Above all, he faults the media for irresponsible hype.

On the positive side is the speed with which the scientific community corrected itself. The establishment erred not on the side of dogmatic rejection but on the side of tolerance. There were no witch-hunts. Deryagin did not settle in Siberia. Perhaps four years of fruitless research and acrimonious debate is a small price to pay for the thorough testing of an honestly claimed anomaly. No one interested in the sociology of science should pass up this absorbing chronicle.

Science in Ancient China

18

Until a few decades ago, most people took it for granted that in science and technology ancient and medieval China had lagged far behind Europe. Then one man, Joseph Needham, slowly, tirelessly, and systematically demolished this myth. At the same time, he uncovered a profound mystery.

Needham is a large, shy, eighty-two-year-old Christian socialist (and practicing Anglican) who began his distinguished career as a Cambridge University biochemist. Every superlative you can think of has been justly applied to his ongoing work, *Science and Civilization in China,* of which eleven big volumes have been published or are on press, with nine yet to come. These incredible tomes leave little doubt that for fourteen centuries China led the world in both pure and applied science.

Science in Traditional China (Harvard University Press, 1981) is a collection of lectures by Needham that forms a pleasant, highly informative introduction to his multivolume major study. Long before other cultures, Needham shows, the Chinese were printing books, using magnetic compasses, controlling insect pests with other insects, making mechanical clocks, casting iron for plowshares and bridges, recording astronomical phenomena, using a decimal system, measuring earthquakes with seismographs, and doing a thousand similar things. No longer can one say patronizingly that the early Chinese used their gunpowder only for fireworks. As the first lecture of Science in Traditional China details, they used explosive powder in guns, bombs, grenades, land mines, flamethrowers, rockets—even for cannon on ships.

Now for the dark mystery mentioned above. Why did China,

This review originally appeared in *Science Digest,* March 1982, and is reprinted here with permission.

with all these achievements and with a philosophical climate more favorable to science than Christianity's fail to make the giant leap into experimental method and modern science that the West did?

This failure cannot be explained away, Needham argues, by reference to China's supposedly fixed attitudes toward time and change. Time in China has always been as objective as stars and stones—not at all like the mind-dependent illusion it tends to be in Buddhism and Hinduism.

It is easy to understand how Needham's love of Chinese culture would at times lead him, as physicist Philip Morrison once wrote, to see European motes while overlooking Chinese beams. Nowhere is this partiality more evident than in his lecture on acupuncture. Needham may be right in believing that this ancient art has great value both as an anesthetic and a cure, and he is certainly right in saying it does not deserve the praise of those who attribute its efficacy to psychic forces. But he goes to considerable lengths to justify acupuncture on shaky grounds—surely the shakiest is that it has lasted so long!—while making only token remarks about why most Western physicians think the practice has little merit beyond its placebo effect.

As for why China did not make the Galilean leap, Needham has yet to find the answer. He ends his little book by suggesting that the solution lies somewhere in "geographical, social and economic conditions and structures which may yet surface to bear the main burden of the explanation."

Great Experiments 19

Great is a word of great vagueness. Selecting twenty great experiments is like selecting twenty great symphonies. Criteria are themselves vague, and no two experts are likely to agree on even half the selections. Nevertheless, Rom Harré, a distinguished Oxford University philosopher of science, has done the job admirably in *Great Scientific Experiments: 20 Experiments that Changed Our View of the World* (Oxford University Press, 1981). Each chapter presents an experiment with such crisp clarity and loving attention to details that the book can be read with huge delight by anyone with the slightest interest in the history of science.

Harré's criteria are hard to fault: experiments that are famous, experiments that significantly advanced the science of their day, and experiments that display remarkable simplicity. His first selection, Aristotle's study of chicken embryos, will surprise those who think of Aristotle as a philosopher who observed nature passively. True, he did not experiment in the modern sense of testing hypotheses with special apparatus, but he did far more than just stare wonderingly at an egg. He actively intervened in nature's workings by the simple method of breaking open chicken eggs at regular intervals during incubation and by cutting up the embryos. His notes reflect that he was hardly an armchair naturalist: "When the egg is now ten days old the chick and all its parts are distinctly visible. The head is still larger than the rest of its body, and the eyes larger than the head, but still devoid of vision. The eyes, if removed about this time, are found to be larger than beans, and black; if the cuticle be peeled off them there is a white and cold liquid inside, quite glittering in the sunlight, but there is no hard substance whatsoever."

This review originally appeared in *Science 83* (September 1983), and is reprinted here, with changes, with permission.

The first section in Harré's book includes experiments that illustrate methodologies—manipulations of variables under controlled conditions—and each tells a dramatic story. I found particularly fascinating the account of how Theodoric of Freibourg in the fourteenth century, a period once considered devoid of empirical science, modeled the drops of a rainbow with a water-filled glass globe to show how the colors form and why the bow is circular. Galileo did not drop weights from the Tower of Pisa, as the story goes, but he did roll bronze balls down polished wooden grooves to show that all bodies fall at the same rate and to derive a formula for their acceleration. Medieval alchemists tried vainly for centuries to transmute one element into another, but it was not until 1919 that Ernest Rutherford in his sensational experiments with radium made the alchemist's dream a reality. He showed that a nitrogen atom disintegrates when bombarded with a heavier radium atom's α-particles and suggested in his conclusion that "we might expect to break down the nuclear structure of many of the lighter atoms." His findings formed the basis for research in nuclear physics.

The famous attempt by Albert Michelson and Edward Morley, his assistant, to measure the Earth's motion relative to a fixed "ether"—a rigid frame of reference supposedly existing throughout the cosmos—concludes the methodology section. The 1887 Michelson-Morley experiment, which failed to find any evidence of an ether, is one that no compiler of great experiments could omit. No experiment with a negative result has had a more revolutionary impact on modern physics. Although Einstein was little influenced by it (he took it for granted that the ether did not exist), the experiment became a cornerstone of relativity theory because it offered a proof for one of Einstein's assumptions. Contrary to what everyone—including the experimenters—expected, light in a vacuum has a constant velocity regardless of the relative speeds of observer and source.

Harré's second section stresses the content of theories, experiments that form and prove new hypotheses. It opens with the discovery of molecular biologists François Jacob and Élie Wollman in 1956 that genes can be transferred from one organism to another. By limiting the number of genes transferred and observing the new behavior patterns of the recipient strain, the French scientists determined the order in which the genes were transferred, a landmark in the history of genetic engineering. This section also includes psychologist James Gibson's experiment on tactile perception. With nothing but a set of humble cookie cutters, he showed that active exploration, not passive reception, is the essential process in perception. We also learn how Antoine Lavoisier discovered oxygen, how Humphry Davy isolated new elements by electrolysis, and how Joseph Thomson discovered the electron. Newton's famous proof that white light contains all colors and Michael Faraday's incomparable investigations of electricity conclude the section.

The final section, emphasizing techniques and instruments, closes with a detailed description of the Stern-Gerlach apparatus, designed by Otto Stern and Walther Gerlach. An ingenious device that led to many fundamental discoveries in quantum mechanics, it uses a magnetic field to split beams of atoms so that basic quantum properties such as angular momentum, orbital plane, and spin can be demonstrated and easily measured.

In addition to providing historical background for each experiment and biographical facts about the scientists involved, Harré also skillfully summarizes later developments that each experiment made possible. Rutherford's work on radioactivity, for example, was such a giant stride toward the release of atomic energy that it enabled H. G. Wells, as early as 1914, to write *The World Set Free,* a novel about how "atomic bombs" would be used in the next world war. Like all philosophers of science, Harré is fully aware of how cultural forces can influence science. But he wisely limits his attention to scientific content as confirmed and modified by later investigations.

Regardless of its revolutions and shifting paradigms, science discloses more and more about the structure of a world not made by us, a world that existed billions of years before any creatures were around to observe it. Harré has little respect for those who perversely deny that scientific knowledge moves steadily closer to truth, even though absolute certainty is never within grasp. This stimulating book documents how cleverly human minds can construct complex instruments for measuring and observing nature and invent wild theories that have a steadily increasing power to explain how the universe behaves and to predict what it will do in the future.

20

Gardner's "Whys" by George Groth

The Whys of a Philosophical Scrivener (Morrow, 1983), by Martin Gardner, is one of the strangest books of philosophical game playing to come along in many a moon. The author seems well acquainted with modern philosophy—indeed, he studied under Rudolf Carnap and even edited one of Carnap's books—yet he defends a point of view so anachronistic, so out of step with current fashion, that, were it not for a plethora of contemporary quotations and citations, his book could almost have been written at the time of Kant, a thinker the author apparently admires.

Gardner is well known for the mathematical-games column he wrote for *Scientific American.* He is also the editor of *The Annotated Alice,* as well as annotated volumes on "The Ancient Mariner," Lewis Carroll's *The Hunting of the Snark,* and a collection of ballads about the mighty Casey who struck out. In addition to his many books about science, pseudoscience, and mathematics and his several children's books, he has also written a curious novel, *The Flight of Peter Fromm.* Disguised as a biography, it chronicles the progressive disillusionment of a young Protestant divinity student at the University of Chicago who, after chucking Christianity, preserves a faith in God. Because the novel's narrator is an atheist, it has been difficult to know whether Gardner's sympathies are with his narrator or with his bewildered student.

Now the secret is out. Gardner's sympathies are not with his narrator. As his new book makes clear, although he has little use for any organized religion, he believes there are good reasons, though only emotional ones, for faith. He is as ruthless as Carnap or Bertrand Russell in dismissing systematic theology as nonsense. An entire chapter is devoted to demolishing proofs of God and poking fun at Mortimer Adler for his unshakable conviction that a valid proof can be formulated. Only an irrational "leap of faith," as Kierkegaard described it, an impulse springing mysteriously from the heart and will, can underpin philosophical theism.

To put it bluntly, Gardner is a simpleminded fideist who sees himself in the tradition of Kant, William James, and Miguel de Unamuno. It is impossible to imagine anyone reading his outrageous confessional (unless the reader is a clone of Gardner) who, however impressed he may be by the author's wide-ranging erudition and rhetorical skill, will not be infuriated by his idiosyncrasies.

The first "why" Gardner asks is why he is a realist—that is, why he believes a mathematically structured universe is "out there," independent of all human minds. "Let me not look aloft and see my own/ Feature and form upon the Judgment-throne." These lines, from a poem by G. K. Chesterton, are the chapter's epigraph. It turns out that Gardner is a fan of G. K.'s, even though he has not the slightest sympathy for Roman Catholic doctrine. He also admires H. G. Wells. Wells and Chesterton? It would be hard to pick two writers more incompatible or about whom today's critics care less. "Can you comprehend," Gardner asks, "as most of my friends cannot, how it is possible to admire . . . the writings of both men? If so, you will understand how it is possible to combine a Chestertonian faith . . . with a Wellsian admiration for science, and at the same time ignore each man's areas of blindness."

After arguing for the reality of an outside world (here Gardner sides with Russell and Hans Reichenbach in making a firm ontological commitment to realism, rather than with Carnap, who defended realism only because he considered it a more efficient language than phenomenology), Gardner takes on the pragmatic theory of truth. In a series of clever arguments based on selecting a card at random from a deck (Gardner is an amateur magician), he concludes that pragmatism failed in its effort to replace the traditional Aristotelian correspondence theory of truth with a theory in which truth is defined as the passing of tests for truth. Although he thinks Russell and John Dewey differed mainly in their choice of language when they clashed repeatedly over this question, he sides strongly with Russell's language. Pragmatism died, Gardner tells us, because the verbal revolution it desired was pragmatically undesirable.

Gardner's chapter on why he is not a "paranormalist" contains little he has not said elsewhere and ad nauseam. He is down on parapsychologists not because he thinks psychic forces are impossible—nothing in science is impossible, he never tires of saying—but because he finds their evidence too feeble beside the wildness of their claims. Would the world be more interesting if psi forces existed? Maybe yes, maybe no. Gardner speculates amusingly on some of the less-pleasant consequences that could result if ESP and PK turn out to be genuine.

In explaining why he is not a relativist with respect to aesthetic values, Gardner goes to preposterous lengths to justify his convictions that "Dante and Shakespeare were better poets than Ella Wheeler Wilcox, that Michelangelo was a greater painter than Jackson Pollock, and that Beethoven's music is superior to that of John Cage or a punk rock band." So what else is new?

There is something to be said for Gardner's defense of objective value judgments in aesthetics, but he spoils it all with a dreary recital of his own peculiar tastes in poetry. No one will fault his admiration for Homer, Virgil, Dante, Shakespeare, Milton, Keats, and Emily Dickinson, but what is one to make of his distaste for Yeats? He considers T. S. Eliot "overrated" and agrees with Nabokov that Ezra Pound was a "total fake." Although he says he has tried his best to enjoy William Carlos Williams, he has yet to find a poem by Williams he thinks worth reading twice. The reader is asked to compare a crude parody of Williams with one of Williams's best-known short poems. Gardner's atrocious spoof—it contains such lines as "Your knees are a southern breeze"—is obviously inferior to Williams's lovely lyric about the butterfly on a red wheelbarrow.

Moral relativism enrages Gardner even more than aesthetic relativism. Here his position is substantially the same as Dewey's: a naturalistic ethics can be based on a common human nature provided one makes such emotional assumptions as that it is better to be healthy than sick and better to be alive than dead. Stale arguments against the extreme cultural relativism that once dominated American anthropology are trotted out and doggedly defended; but when it comes to the "staggering" moral decisions that will have to be made when biologists find ways to alter human nature, Gardner writes, "I have no light to throw on these rapidly approaching and terrible questions."

Free will is the next topic to occupy Gardner's attention. No modern philosopher is likely to be impressed by his simple way of evading this ancient conundrum. He "solves" it by declaring it unsolvable. As Gardner sees it, the fundamental dilemma is that determinism leads straight to fatalism but that indeterminism is even worse because it turns free will into the haphazard toss of a die inside one's skull. There is no way, he insists, to define free will

without sliding into one or the other of these dark chasms. The best
we can do, indeed the *only* thing we can do, is leave will a blinding
mystery. It is not fate, it is not chance. It is somehow both, yet
somehow neither. "Ask not how it works," he concludes, "because
no one on earth can tell you."

When he comes to politics and economics, Gardner's chaotic
high jinks seem calculated to drive both liberals and conservatives
up the wall. Gardner has no respect for what he calls the "Smithi-
ans"—all those who think government should shrink, leaving as
much play as possible for the free market. Robert Nozick's minimal
state is dismissed as a "me generation" aberration, and Ayn Rand
is shoved aside as the ugly offspring of Milton Friedman and Mada-
lyn Murray O'Hair. Withering scorn is heaped on the supply-siders.
He quotes Paul Samuelson's remark that if Friedman did not exist it
would be necessary to invent him. He likens Friedman to a chiropractor. Unlike an authentic doctor who knows too much to make
a snap diagnosis, a chiropractor will tell you at once why your back
is aching and how quickly he can cure it.

On the other hand, conservatives will be delighted by Gardner's jabs at Karl Marx. He quotes an amusing passage from a forgotten book on Russia by Wells in which the insane abundance of
Das Kapital is compared with Marx's woolly beard. The sooner Michael Harrington forgets about Marx, says Gardner, the better. Polit-
ically, Gardner turns out to be—who could have guessed it?—an
old-fashioned democratic socialist in the tradition of Wells, Russell,
Norman Thomas, Gunnar Myrdal, Irving Howe, and a host of other
socialists who are as ignored today by most liberals as they are
hated by all conservatives—and whose practical political prospects,
which he does not discuss, seem as dim as ever.

We are now halfway through Gardner's bizarre book—and
ready for its biggest surprise, his back flip into fideism. But first
he writes a diversionary chapter on polytheism. Like Lord Dunsany—whose name suggests how out of date are Gardner's tastes but
whose fantasies he admires—Gardner has a wistful fondness for the
beautiful gods of ancient Greece and little to say about their cruelty.
Although he finally chooses monotheism, it is largely on the flimsy
ground of "Occam's razor." Emotionally, he believes, a single God
will do all a plurality would—and do it better—though in a final
sense he says he does not know whether God is one or many, or
even whether numbers have any meaning when applied to God. He
sees Christianity as almost as polytheistic as Hinduism. Are not Jesus and the Holy Ghost (not to forget Satan, the Immaculate Mary,
and the vast medieval hierarchy of angels) manifestations of a

165

higher deity—just as Brahma, Vishnu, and Siva are manifestations of Brahman?

At this point one might expect Gardner to glide into a pantheism along the lines of Alfred North Whitehead's, but no. He dislikes pantheism even more than polytheism. His God is "personal," though he emphasizes, with Thomas Aquinas and Charles Peirce, that we have not the foggiest notion of what it means to pin human traits on God. He applauds today's Christian feminists for their attacks on the male bias of the Bible but goes them one better. He sees no way the bias can be removed without removing the Incarnation itself and therefore abandoning the heart of Christianity. Nevertheless, if God is to be of any value to us we must model him with the highest metaphors we have. Gardner quotes a colorful passage from C. S. Lewis on what happens when God is modeled with nonpersonal symbols. He becomes a kind of gas—or maybe jello—that permeates the cosmos, of less use to us than a cloud or stone.

There are more surprises. Not only does Gardner believe in God; he also believes that petitionary prayer can make a difference. How? He does not know. As for skeptics, "Do they think, the fools," Gardner quotes from Thornton Wilder's *The Cabala,* "that their powers of observation are cleverer than the devices of a god?" For Gardner, the mystery of prayer is bound up with the terrible mysteries of time, causality, and free will. To defend the right to pray, he constructs several ingenious models, one of them deriving from quantum mechanics. They are put forth whimsically. His only motive, he claims, is to show that belief in the efficacy of prayer is not logically contradictory. Are any of the models true? "Do not ask *me,*" Gardner answers himself.

One of the characteristics of Gardner's "theological positivism," as he calls it, is that he is content to accept paradox and mystery in regions where philosophers endlessly seek solutions. For a theist, the most dreadful of all mysteries is random, insane evil. Two chapters are devoted to the ancient argument that God either (a) could prevent evil but does not and hence is not good or (b) wants to prevent it but cannot, in which case He is not all powerful. Gardner not only has no answer to this deadly dilemma; he actually thinks it makes atheism more sensible than theism! All the better arguments, he freely admits, are on the atheist's side. The leap of faith is an irrational, absurd somersault of the soul that some people cannot avoid making (Gardner does not know why) even though all experience suggests that the leap is as foolish as Don Quixote's belief that Dulcinea smells like sweet perfume. The modern fideist, Gardner writes, must grant it all.

Note how Gardner.here ensures that no one can prove him wrong. His invisible God is like the White Knight's green whiskers; no one can see them because he keeps them always behind his fan. The atheist argument from evil to no God bounces harmlessly off Gardner's head because he does not deny its persuasiveness. Like Pascal, he defends his fideism on the grounds that if it were otherwise, if we knew the secret of evil, faith would not be faith. It would become compelled belief.

What is one to say about such a view, wholly unsupported by reason or revelation? I can best reply with a passage from Russell that Gardner must know but apparently could not bring himself to quote: "There is to my mind something pusillanimous and sniveling about this point of view, which makes me scarcely able to consider it with patience. To refuse to face facts merely because they are unpleasant is considered the mark of a weak character, except in the sphere of religion. I do not see how it can be ignoble to yield to the tyranny of fear in all ordinary terrestrial matters, but noble and virtuous to do exactly the same thing when God and the future life are concerned."[1]

Gardner's discussion of immortality is the most outlandish in the book. Although he realizes that within one's head theism can be separated from hope for another life, he follows Unamuno in regarding the two beliefs as interlocked inside the heart. He quotes Unamuno's conversation with the peasant who, after being told that perhaps there is a God but no afterlife, responded, "Then wherefore God?" Although Gardner believes Jesus to have been an ordinary man, likely born illegitimate and possibly gay, he professes to admire most of what he suspects Jesus actually taught. He is amazed that Paul Tillich, who did not believe in a personal God or an afterlife—Jesus' two basic themes—could have made the cover of *Time* as a great Christian theologian. As for hell, which Gardner thinks Jesus also taught, he cites this as one reason why he stopped calling himself a Christian.

Gardner constructs three models for an afterlife, all designed (like his models for prayer) to show that the doctrine is not logically inconsistent. Is one of the models true? "For my part," Gardner answers, "I believe that none of the models . . . is true. I am persuaded that the truth about immortality is as far beyond our grasp

---

1. From *The Value of Free Thought: How to Become a Truth-Seeker and Break the Chains of Mental Slavery,* by Bertrand Russell (1944). Reprinted in *Understanding History and Other Essays* (Philosophical Library, 1957).

as the ideas in this book are beyond the grasp of a glowworm." Again it is all a matter of "faith," for which he can show no rational basis.

The book's last chapter but one is a frank attempt to arouse in the reader a sense of what Rudolf Otto called the "numinous," a Chestertonian awe before the incredible mystery of existence. The final chapter pleads for religious tolerance. Gardner is appalled by the view that history is a duel to the death between Christianity and atheism—a duel that Chesterton and Whittaker Chambers saw, and William Buckley and Ronald Reagan still see, as manifest today in the military confrontation of "Christian" America and "atheist" Russia. Gardner quotes a poem by Stephen Crane about a "complacent fat man" who climbed to the top of a mountain, expecting to see "good white lands and bad black lands," only to find that the scene was gray. This leads to the book's final metaphor. Today's philosophical grayness becomes a backdrop that intensifies the colors of an unpredictable future.

How seriously should we take Gardner's fideism? He seems sincere, yet one wonders. After all, the man has a reputation as a hoaxer. His April 1975 column in *Scientific American* purported to disclose such dramatic breakthroughs as the discovery of a map that required five colors, a fatal flaw in relativity theory, an opening move in chess (pawn to queen's rook four) that is a certain win for white, and a lost parchment proving that Leonardo da Vinci invented the flush toilet. Thousands of readers wrote to tell Gardner where he went wrong, and one irate professor tried to have him expelled from the American Mathematical Society. Happily, the society made him an honorary life member. George Groth, by the way, is one of Gardner's pseudonyms.

21

How Science Self-Corrects

Why do scientists treat with respect such wild concepts as quarks and black holes but dismiss as nonsense extraterrestrial UFOs and the cosmology of Velikovsky? This is one of the central questions raised by Richard Morris, a physicist turned science writer, in *Dismantling The Universe* (Simon and Schuster, 1983), an admirable popular account of how science operates by dismantling faulty theories and replacing them with better ones. The process may never end and final truth may be unreachable, yet who can deny the fantastic success of the enterprise?

Morris begins with an excellent nontechnical summary of how relativity theory, in spite of its wrench to common sense, improved on Newton's physics. Compared with the slow swing in earlier centuries, from Earth-centered cosmologies to heliocentric models, the relativity revolution was amazingly rapid and painless. And quantum mechanics, the next great revolution, was even swifter.

Morris has little enthusiasm for Fritjof Capra and other physicists who contend that quantum mechanics supports Eastern mysticism. He knows that Erwin Schrödinger, one of quantum mechanics' architects, was profoundly interested in Eastern philosophy and that Niels Bohr saw the yin-yang design as a symbol for complementarity—the view that seemingly contradictory aspects of quanta are opposite sides of an incomprehensible truth. But it takes a lot of distortion, Morris believes, to find quantum mechanics closer to Eastern than to Western thinking. Eastern sages have no monopoly on interconnectedness. Western mystics had similar visions of a supernal Oneness, not to mention the long lines of European pantheists from Plotinus to Alfred North Whitehead. It was not a Taoist but

This review originally appeared in *Science 84,* March 1984, and is reprinted here with permission.

nineteenth-century English poet Francis Thompson who wrote: "Thou canst not stir a flower/Without troubling of a star."

A chapter on the pitfalls of intuition tells how Copernicus, having no way to know that planetary orbits are elliptical, was forced to use epicycles as ugly and complicated as those of the Ptolemaic models. So entrenched was the belief that orbits must be circular that even Galileo refused to accept Kepler's ellipses. Kepler's intuition led him far astray when, in a burst of bogus illumination, he related the orbits of the known planets to the five Platonic solids. Einstein's remarkable intuition failed him utterly when he introduced a "cosmological constant"—a repulsive force between particles—to prevent gravity from collapsing his elegant steady-state model of the universe.

Of course none of these mistakes detracts from the monumental achievements of the three men. For Morris, they highlight the fumbling way that guesses in science are made. When it became apparent that the universe is expanding, Einstein lost no time in calling his constant "the chief blunder of my life." Morris is surely right in seeing this corrective process as operating with increasing efficiency as science more and more becomes a vast cooperative enterprise with rapid exchanges of information between its various research centers.

A hard-hitting chapter on crackpottery should be read by anyone who thinks the term useless. Velikovsky, for example, in *Worlds in Collision*—in which he theorized that Venus was originally a comet expelled from the planet Jupiter that eventually collided with Mars and Earth—did not really challenge anything. "On the contrary," Morris says, "he behaved as though modern science did not exist." In Morris's opinion, now shared by everybody except a few tiresome diehards, Velikovsky's fantasies were so far outside the bounds that he became the very model of a crackpot.

The term should not, however, Morris cautions, be indiscriminately applied to research programs merely because they prove to be mistaken. Consider polywater, a strange jellylike water that excited chemists in the mid-1960s; Vulcan, a tiny planet once thought to be inside Mercury's orbit; and the fictitious N-rays that some French physicists at Nancy believed they had observed. Were the defenders of these notions crackpots? Morris thinks not. The science community became skeptical, he points out, "when these theories became so convoluted that they began to take on crackpot proportions." Unnecessary insults may have been tossed back and forth, but the speed with which these three controversies vanished are striking tributes, Morris contends, to science's constantly improving skill in self-correction.

Is it possible for a theory to be reputable yet so bizarre that no one can believe it? Morris introduces two recent instances: the many-worlds interpretation of quantum mechanics, in which the cosmos fractures every microsecond into billions of parallel worlds, and a speculation of physicists Robert Dicke and P. J .E. Peebles that there are countless universes within universes, each of them expanding from an exploding black hole. That establishment journals regularly publish such outlandish but possibly useful ideas surely belies the view that mainstream science is a rigid orthodoxy, contemptuous of off-trail thinking.

Morris ends his book by wondering just what is meant by *real*. The question is unusually troublesome on the subatomic level, where particles seem not to exist until they are measured. Are the particles no more than figments of human imagination? In an obvious sense they are mental constructs, yet in another sense, Morris wisely insists, they model something out there, independent of human life, that is sufficiently structured to impose severe restraints on the modeling.

What roles do simplicity and beauty play in the invention of a good hypothesis? Here I think Morris is right in general, but in spite of Keats's famous line about truth and beauty, I must question the book's final sentence: "Science seeks to create pictures of the order in nature which are so logically elegant that we cannot doubt that they are true." No one knows how to measure a theory's elegance— that mysterious mix of simplicity and beauty. It is a theory's success in passing new empirical tests that must always be the final arbiter. Particle physicists now like to say that the recent theory of supersymmetry is much too beautiful to be false. Alas, this has been said many times before about theories that bit the dust because, though very beautiful, they were also very wrong.

22   Some Trends in Mathematics

Michael Guillen, a science writer with a doctorate in physics, once taught a course at Cornell University called "Mathematics for Poets." His *Bridges to Infinity* (Tarcher, 1983), subtitled "The Human Side of Mathematics," is similar in intent: to introduce readers who are fearful of mathematics to some of its beauty and wonders.

He has done the job well. In brief, brisk, entertaining chapters he offers startling glimpses into such forbidding topics as limits, imaginary numbers, singularities, groups, non-Euclidean geometry, probability, game theory, and topology. A chapter on other dimensions touches on the fantastic fractal curves of Benoit Mandelbrot which have been astounding the science community in recent years and providing eerie landscapes for such films as *Star Trek II*. Catastrophe theory, a mathematical fad a few years back, gets another chapter.

The book's final topic, combinatorics, includes the currently hot research area of "computational complexity," where problems are encountered that are solvable only by running a computer for millions of years. The notorious "traveling-salesman problem," for example, asks for the shortest route that visits n points on the plane. Computers can handle this when n is small, but, as n increases, running time rapidly accelerates to impractical lengths. It belongs to a category known as NP complete. These are problems interlocked in such a way that if a procedure is ever found for solving one of them in manageable time, all the others will be solved.

At the heart of Guillen's book is a chapter on Kurt Gödel's famous proof that formal systems complex enough to include arithmetic contain theorems that cannot be proved within the system.

This review originally appeared in *Book World,* 4 March 1984, and is reprinted here with permission. © 1984 by The Washington Post.

Guillen shares the conviction of a small but vocal group of mathematicians that somehow this makes mathematics as uncertain as empirical science.

I cannot agree. Gödel did indeed show that every formal system beyond a certain level of complexity must contain theorems impossible to prove or falsify without going to a larger system which in turn will harbor undecidable statements. But from this it does not follow that all theorems are undecidable. In fact, one has to look hard for interesting theorems that may be undecidable, such as Guillen's example: Goldbach's conjecture that every even number except 2 is the sum of two primes. The four-color map theorem was long considered a promising candidate for undecidability in topology, until computers recently decided it.

Imre Lakatos was a Hungarian philosopher noted for his vigorous efforts to blur the distinction between mathematical proof and empirical corroboration. Guillen is such an admirer of Lakatos that in the chronology at the back of his book he places him among twenty-two eminent mathematicians, since the year one, who are mentioned in the book. But Lakatos made no significant contributions to math. He was a philosopher of science, much less important in the history of mathematics than John Von Neumann and others who figure prominently in the book but fail to make the chronology.

Mathematics has always been saturated with uncertainty. The only change has been that now we know, as earlier mathematicians did not, that in principle mathematics will never be free of uncertainty. In this respect mathematics does resemble empirical science. This does not mean, however, that there are reasons for doubting 17 is a prime or that the Pythagorean theorem cannot be proved in a manner qualitatively unlike, say, the way quantum mechanics is supported by experiments. Given the axioms and rules of a formal system such as elementary Euclidean geometry, theorems have the same kind of certainty as the assertion that a yard equals three feet. Their proofs are in no way contravened by the existence of non-Euclidean geometries or by theorems undecidable within the Euclidean system.

The book contains some mistakes, such as making 1 a prime and calling Von Neumann a German (he was by birth Hungarian). And Guillen has a tendency to end chapters with philosophical asides that I find murky. Examples: Topologists are said to be better than other mathematicians "in their understanding of soul," and our mental "quirks" are "analogous to topological invariants." Is anything gained by likening the human mind to the null set (the set with no members) because John Conway constructed infinite families of surreal numbers by applying simple rules to the null set? What does it mean to say the mind contains infinity and is a singularity? Why does Guillen wonder whether catastrophe theory would hold on other inhabited planets, when obviously it must?

I am unable to follow Guillen's argument that if other planets are inhabited, the greater the reasoning power of the aliens, the less interested they would be in games. The reason seems to be that any intellectual game becomes dull to anyone bright enough to analyze it completely. As a good Gödelian, Guillen should expect superbeings to play supercomplex games with the same enjoyment that we play bridge and chess. And what on earth can Guillen mean when he applies NP completeness to human behavior, suggesting that if we can solve a "single key problem" in the social sciences, it might follow that "all problems related to it will thereupon be resolved"?

Perhaps these caveats reflect my own biases. In any case, the book's sparkling miniessays can be read with as much delight by nonmathematicians as nonpoets can relish the poems in a good anthology.

Comfort's Comforts

Dr. Alex Comfort, trained in classics at Cambridge University, is the British gerontologist who made a fortune with his best sellers *The Joy of Sex* and *More Joy of Sex*. Discussions with bright students at the Neuropsychiatric Institute, University of California at Los Angeles, where Comfort now teaches, impelled him to set down on paper the ideas that have swirled in his brain since he discovered the joy of QM (quantum mechanics). The result is *Reality and Empathy: Physics, Mind and Science in the 21st Century* (State University of New York Press, 1984).

Comfort's main theme, like that of many another recent book, is that QM is such a revolutionary new way of looking at the universe that if it could be "popularly empathized, it would be a blockbuster" (P. 25). By "empathize" Comfort means made so intuitively clear that nonphysicists could *feel* what it is all about.

The book's first sentence, "Worlds are created by brains," is intentionally ambiguous. You quickly learn that "world" means a world model or map of reality, but later on you encounter the view that perhaps brains have created the outside world as well. Objective reality and mind could be related by what Douglas Hofstadter calls a "strange loop." Somehow—just how is the ultimate mystery—Being was able to bifurcate into matter and minds that allow the matter to see itself. Matter and mind may be epiphenomena of one another, like Escher's picture of two hands, each sketching the other.

Comfort professes to be immune to the epidemic of "pseudo-east nonsense" now infecting the West—what he calls a mix of "Aquarians, acid-heads, and amateur mystics," speaking "yoga-

This review originally appeared in *Nature,* 26 April 1984, and is reprinted here, with changes, with permission.

babble" and tossing up "premature Taoists who write popular books on physics" (Pp. 37–38). The public would do well to stop listening to "itinerant swamis who preach in mottos out of fortune cookies" (P. 33) and to turn instead to the original sacred literature of Buddhism and Hinduism, where they would find that introspection had indeed produced visions in surprising harmony with the empirical results of modern physics.

What do QM and Eastern thought have in common? Comfort believes it is a way of seeing our pluralistic phenomenal world as an illusion produced by an impenetrable, timeless, unfathomable reality. Comfort seldom calls this reality God, preferring instead the impersonal Brahman of Hinduism. Although QM makes no ontological statements, it is nevertheless saturated with anomalies that Comfort thinks support this Eastern insight.

Consider the notorious EPR (Einstein-Podolsky-Rosen) paradox, invented by Einstein and two associates as a thought experiment but which recently has received strong support from laboratory experiments. Two photons (in one version) are emitted in opposite directions by an interaction that gives them opposing spins. In QM, neither particle has a definite spin until measured, yet the two are so correlated, that if you measure A, thereby creating, say, a plus spin, B will acquire a minus spin even though it may be light years away.

Einstein believed that his paradox proved the incompleteness of QM. Comfort agrees. The best way out, he thinks, is to adopt what he calls the "thingless universe" of David Bohm, a QM expert who has long been an admirer of Eastern philosophy. In Bohm's vision, particles are "explicates" of an "implicate order," a substrate that is outside our space and time. Comfort likens the particles to spots that seem to move on the screen of a computer game. But nothing is really moving. Points of light are merely switching on and off in obedience to signals from invisible hardware. Perhaps Zeno was right. Motion is unreal. The outside world is what the Hindus call *maya,* an illusion conjured up by the motionless Brahman.

It is easy to see how this vision could furnish support for the psi forces that parapsychologists claim are indifferent to space and time. Comfort makes a great pother about his neutrality with respect to psi: "I have no idea whether paranormal phenomena exist or not" (P. 229) However, he is a vigorous advocate of what he calls "demonic" conjectures—efforts to see the world in "non-human" ways. It may be, he writes, that altered states of consciousness give authentic glimpses into bizarre but fruitful world models. He shares with psychoanalyst Jan Ehrenwald, an ardent advocate of psi, the conviction that anecdotal evidence for psi, such as telepathic dreams, is far stronger than any laboratory result. Such evidence is so voluminous, Comfort says, that to attribute all of it to self-deception, coincidence, or fraud seems to him like doubting the existence of badgers.

Scientists who became Spiritualists, such as Oliver Lodge, Comfort considers much less credulous than skeptics suppose. They "probably" did not see discarnate spirits, he allows, though they "apparently" did observe "uncanonical transfer of information" (P. 218). These are curious remarks coming from a man who insists that his personal estimate of the odds for psi being genuine are fifty-fifty.

Comfort's fondness for demonic models prompts him to look benignly on many other conjectures that most scientists consider hogwash. Rupert Sheldrake, for instance, is persuaded that members of a species are united by a "morphogenetic field." If you train some rats to find their way through a maze at Harvard University, rats of the same species will learn to run the maze faster in Scotland. Sheldrake may be wrong, Comfort admits, but he is raising an important question. Comfort is pleased that Sheldrake's challenge has not affected colleagues like an "inopportune flatus in an elevator."

Karl Pribram's holographic model of the brain is another demonic theory that Comfort finds comforting. The universe may be a monstrous hologram, each tiny part, like one of Leibniz's monads, containing the whole. The book is not yet closed, Comfort is also convinced, on the role of Lamarckism in evolution. His wildest speculation is that sabre-toothed tigers, which flourished before humanity was on the scene, may not have really been "there" except in a vague way, their pale reality sustained only by the low-order brains of the beasts that saw them!

Comfort borrows from Hofstadter the whimsy of interrupting his prose with comic dialogues. A lion and a unicorn step down from a coat of arms to argue about scientific method. A snake named Wilberforce, after the cleric who debated with T. H. Huxley, discusses evolution with a mockingbird. Gezumpstein, a demon from beyond spacetime—his sole task is to invent testable hypotheses—models reincarnation with a row of isolated spots created by painting a line along one side of a helix. Adam demands of God, his psychiatrist, that she reveal the real reason why he was kicked out of Eden.

The cleverest of these interludes tells how Gezumpstein's conjectures take the form of balloons. He distributes them to scientists who blow them up and keep them inflated until they are punctured by a fact. The facts are called "poppers", a play on the name of Karl Popper. Balloons given to mathematicians last the longest, but no balloon is "popper-proof." Many last for centuries before bursting. Some are allowed to deflate, only later to be blown up again.

The book is stimulating, funny, quirky and marred by a rambling, repetitious, disjointed organization. Comfort has a fondness for awkward terms such as "homuncularity," "pre-scientoid," and "dogsbody," a neologism borrowed from James Joyce's *Ulysses*. He seems to have no interest in any modern Western philosopher except Popper. George Berkeley, who more than any other thinker struggled with all of Comfort's ontological puzzles, is not in the index, though I discovered a trivial reference to him on page 197.

Readers unacquainted with modern physics will find most of the book unintelligible. A section on how each moment of history can be undetermined, even though Brahman is timeless and unchanging—it just *is*—I found totally opaque. And although I learned much, I finished the book with a dizziness from its endless zigzags and with a feeling that Comfort could have made his points much clearer if he had tried.

Calculating Prodigies

In Brunswick, Germany, in 1780, a stonemason was calculating the wages due his workmen at the end of the week. Watching was his three-year-old son. "Father," said the child, "the reckoning is wrong." The boy gave a different total which, to everyone's surprise, was correct. No one had taught the lad any arithmetic. The father had hoped his son would become a bricklayer, but thanks to his mother's encouragement, the boy, Carl Friedrich Gauss, became one of the greatest mathematicians in history.

Regardless of such anecdotes, the ability to calculate swiftly and accurately in one's head seems to have little correlation with creative mathematical ability or high intelligence. Some eminent mathematicians—Gauss, John Wallis, Leonhard Euler, and John von Neumann, to name four—had this ability, but most first-rate mathematicians were and are no more skilled in mental arithmetic than are good accountants. A few calculating prodigies have even been mentally retarded. No one knows the extent to which this curious skill is genetic or how much is the result of environment and arduous self-training.

Steven B. Smith, in his admirable book *The Great Mental Calculators* (Columbia University Press, 1983)—the best, most comprehensive, most reliable book yet written on the subject—thinks that the talent springs mainly from strong childhood motivations. For a variety of reasons not well understood, a child, often one who is isolated and lonely, will fall passionately in love with numbers. "Children need friends," Smith writes, "for amusement and companionship. They often devise imaginary friends to keep them company when flesh and blood friends are absent. Calculating prodigies have made numbers their friends."

When you and I see a license plate on a car ahead of us, we usually see a meaningless number, but to calculating geniuses it is invariably rich in properties and associations. If it happens to be prime (a number with no factors except itself and 1), they will instantly recognize it as a prime. If it is composite (nonprime), they may at once determine its factors. Consider 3,844. "For you it's just a three and an eight and a four and a four," said William Klein to Smith, who considers Klein the world's greatest living mental calculator. "But I say, 'Hi, 62 squared.'"

There is an interesting parallel, Smith suggests, between mental calculating and juggling. Almost anyone can learn to juggle, but only a few are driven to practice until they become experts, and even fewer make it their profession. Children who learn to juggle numbers in their head diverge in later life along similar paths. Some lose interest in the art, some preserve it as a hobby, some make good use of it in their careers. On rare occasions, when talent and passion are high and environmental influences appropriate, a young man or woman will work up a "lightning calculation" act and go into show business.

The stage calculators—like magicians, jugglers, acrobats, tap dancers, chess grandmasters, and pool hustlers—are a diversified breed, with almost nothing in common except their extraordinary ability. Consider Zerah Colburn, one of the earliest and fastest of the calculating wizards. Born in 1804, the son of a poor Vermont farmer, he was only six when his father began to exhibit him. Within a few years he became a celebrity both here and abroad. Washington Irving helped raise money to send Zerah to school in Paris and London. After his education, he gave up his stage career to become a Methodist preacher.

Before Colburn died at age thirty-five, he wrote a quaint autobiography with the title *A Memoir of Zerah Colburn; written by himself—containing an account of the first discovery of his remarkable powers; his travels in America and residence in Europe; a history of the various plans devised for his patronage; his return to this country, and the causes which led him to his present profession; with his peculiar methods of calculation.* Aside from his calculating prowess, the only other notable aspect of the man was that he was born with six fingers on each hand and six toes on each foot, like the giant of Gath (with whom he felt a kinship) mentioned in the Old Testament (I Chron. 20:6).

The chapter on Colburn is one of the most fascinating in

Smith's book. The rustic youth's methods were discovered by himself, but unlike many stage calculators he did not mind explaining them. When the Duke of Gloucester asked how he had so quickly obtained the product of 21,734 and 543, Colburn said he knew at once that 543 was three times 181. Because it was much easier to multiply by 181 than by 543, he first multiplied 21,734 by three, then multiplied the result by 181.

While Colburn was becoming famous in America, his counterpart in England, George Parker Bidder, was on tour as a nine-year-old calculating prodigy. The two boy wonders eventually met for a contest in Derbyshire (Colburn was fourteen, Bidder twelve), but there was no clear victor. After an education at the University of Edinburgh, Bidder became a first-rate civil engineer, retaining his calculating powers throughout a long, happy, and productive life.

Shortly before he died, Smith tells us, Bidder was visited by a minister who had a strong interest in mathematics. From the nature of light, Bidder told his guest, one could gain insight into both the largeness and the smallness of the universe's structure. Light travels, he said, at 190,000 miles per second (the best estimate of the day), yet space is so vast that it takes light an enormous time to go from star to star. At the other end of the scale, the wavelength of red light is so small that 36,918 waves extend only an inch. The minister wondered how many waves of red light would strike the eye's retina in one second. "You need not work it," said Bidder. "The number of vibrations will be 444,433,651,200,000."

Among this century's professional mathematicians, the greatest all-around mental calculator was Alexander Craig Aitken, professor at the University of Edinburgh and the author of several textbooks and some eighty papers. His lecture in 1954, "The Art of Mental Calculation," is the richest source in print on the psychological processes involved in the art.

As Aitken said in his speech, and as Smith stresses in his book, calculating prodigies fall roughly into two groups: those who "see" numbers in their minds and those who "hear" them. Auditory calculators, such as the Dutchman Klein who can multiply any two ten-digit numbers in about two minutes, usually accompany their mental labors with muttering or at least lip movements. Visual calculators stare silently at the written numerals or off into space. Aitken could not decide whether he was visual or auditory. Here is how he put it:

Mostly it is as if they [the numbers] were hidden under some medium, though being moved about with decisive exactness in regard to order and ranging; I am aware in particular that redundant zeros, at the beginning or at the end of numbers, never occur intermediately. But I think that it is neither seeing nor hearing; it is a compound faculty of which I have nowhere seen an adequate description; though for that matter neither musical memorization nor musical composition in the mental sense have been adequately described either. I have noticed also at times that the mind has anticipated the will; I have had an answer before I even wished to do the calculation; I have checked it, and am always surprised that it is correct.

Like all lightning calculators, Aitken had a prodigious ability to memorize long strings of digits. There are ways to do this by clever mnemonic systems—translating groups of digits into picturesque words, then joining the words by outlandish images—but such techniques are much too slow for rapid calculators. In doing mental multiplication, for example, partial products have to be fixed in the mind until the process is completed in just a few seconds. Aitken mentioned in his lecture that he once amused himself by memorizing pi to a thousand decimal places. Smith tells how Aitken interrupted his talk to recite this long chain of patternless digits, while someone checked its accuracy—there were no errors—against a table of pi. Hans Eberstark, an Austrian mental calculator, has memorized pi to more than 10,000 places, and Smith cites others who have gone far beyond that.

Aitken distrusted all mnemonic tricks. "They merely perturb with alien and irrelevant association a faculty that should be pure and limpid." His way of memorizing pi was to arrange the digits in rows of fifty each, then divide each fifty into ten groups of five and "read these off in a particular rhythm. It would have been a reprehensibly useless feat, had it not been so easy." Interest in the sequence makes the task of memorizing much easier, Aitken added. "A random sequence of numbers, of no arithmetical or mathematical significance, would repel me. Were it necessary to memorize them, one might do so, but against the grain."

In the demonstrations of calculators such as Aitken who are not in show business, there is no need to deceive, but when we turn to the acts of the vaudevillians, where the purpose is to astound and entertain, there are overwhelming temptations to use subterfuges that make the show even more impressive. Smith covers them all. Suppose, for instance, a stage performer has asked his audience for two five-digit numbers. He may say, "Will you please repeat that last number? I'm not sure I heard it correctly." While saying this, and while the number is being repeated, he has already started multiplying in his head.

The performer may slowly chalk the two numbers on a blackboard. By the time this is finished, additional seconds for calculation have been gained. A reporter seeing him write the product immediately after writing the two numbers will understandably be convinced that the performer obtained the sum in two seconds, not realizing that the actual calculating time was closer to two minutes. Moreover, the best methods of multiplying large numbers in the mind build up the product from left to right. This gives the performer still more time. While he is writing the product left to right, he is continuing to calculate digits near the end.

Another important secret of lightning calculation, though stage performers often falsely deny it, is that early in life they have memorized the multiplication table through 100. Thus, in operating with large numbers, they can handle the digits by adjacent pairs and cut calculating time in half. Moreover, thousands of large numbers of the sort that come up often in audience questions—such as the number of seconds in a year or inches in a mile or the repeating sequence of digits in the decimal form of 1/97 (it has a period of ninety-six decimals)—can be committed to memory. No stage performer would ever say: "Please don't ask me that, because I already know the answer."

Some stage calculators are not beneath planting confederates in the audience to call out problems for which the answer is already known. Smith thinks this occurs rarely, but I am not so sure. I know many magicians who do what the trade calls a "mental act"—feats that purport to be accomplished by psychic powers. Their use of secret accomplices is quite common. Why should show-biz calculators be less deceptive?

In some lightning-calculation tricks, the use of an accomplice is cleverly hidden by the fact that only part of a problem need come from a confederate. A marvelous example of this is provided by Smith's account of a performance in 1904 at the University of Indiana by a calculator who called himself "Marvelous Griffith." Griffith wrote 142,857,143, on the blackboard, a number probably called out by a confederate. A second nine-digit number was then supplied by a professor whom everyone knew could not be an accomplice. While the professor was still writing his number, Griffith began chalking the product of the two numbers from left to right. When it was found to be correct, the audience stood up and cheered.

As Smith points out, although Griffith was a skilled mental calculator, this feat can be done by anybody. You have only to divide the second (legitimate) number by seven, and do this twice. If there is a remainder after the first division, carry it back to the initial digit and divide through again. If the final division does not come out even, you goofed. The trick works because 142,857,143 is the quotient when 1,000,000,001 is divided by seven.

Unfortunately, the numbers involved in this little-known swindle are beyond the capacity of pocket calculators; but if you want to astound your friends, there are simpler tricks based on the same principle. For example, the product of 1,667 and any three-digit number *abc* can be obtained mentally as follows. Add zero to the end of *abc,* then divide it by six. If there is no remainder, continue by dividing *abc* by three. If the remainder after the first division is two or four (it cannot be anything else), carry half the remainder (one or two) back to the first digit and divide *abc* by three.[1]

Most stage performers include in their act two demonstrations that are much easier than they seem: giving the roots of perfect powers and giving the day of the week for any date called out. Only small charts need be memorized for finding cube roots (they are easier to calculate than square roots), and fifth roots are still simpler because their last digit is always the same as the last digit of their fifth power. You will find the details in Smith's book, as well as an excellent system (there are many) for doing the calendar trick. Both feats can be easily mastered by anyone who cares to spend a little time practicing.

Jugglers, by the way, also are not beyond augmenting their skill with fakery. Because showmanship is the essence of good vaudeville, one can forgive professional jugglers and acrobats and calculators for highlighting their acts with harmless flimflam, but it means that in reading accounts of their marvels one must often take them with a grain of salt. You do not have to believe that Unus, a circus acrobat, can actually do a handstand on one gloved finger even if you see him appear to do it. You do not have to believe, when you read in the 1984 *Guinness Book of World Records* that Shakuntala Devi of India (one of the few women calculators currently performing) multiplied two thirteen-digit numbers, each randomly selected by a computer, in twenty-eight seconds. As Smith politely understates it, "such a time is so far superior to anything previously reported that it can only be described as unbelievable."

Do lightning-calculation acts have a future, or has the computer rendered them uninteresting? At the close of his lecture, Aitken admitted that his own abilities began to decline when he got

1. For other magic numbers of this sort and an explanation of why they work, see the chapter on lightning-calculation tricks in my *Mathematical Carnival* (Knopf, 1975).

his first desk computer. "Mental calculators, then, may, like the Tasmanian. . . . be doomed to extinction," he said. "You may be able to feel an almost anthropological interest in surveying a curious specimen, and some of my auditors here may be able to say in the year 2000, 'Yes, I knew one such.'"

Perhaps Aitken was right. On the other hand, the electronic calculator has in certain respects increased the entertainment value of such shows. One of the things that slowed up past performances was the inordinate amount of time required to verify large calculations. Frequently persons in the audience would offer a problem they had previously calculated incorrectly, and much time would be lost in setting the record straight. Today, mental calculations with big numbers can be verified quickly by anyone with a small computer. As young Arthur Benjamin has discovered—he is the only American now performing a mental-calculation act—this makes possible many entertaining feats that were not available to earlier performers.

The use of computers also raises a truly dreadful possibility. So far as I know, it has not yet been exploited on the stage, though I would expect it to be eventually. There are now simple devices for wireless communication between two persons, with receivers tiny enough to be concealed in the ear or the anus to provide easily heard or felt beeps. There is nothing to prevent a confederate backstage—or even sitting in the audience—from quickly solving a problem on a computer, then secretly relaying the answer to a performer, who may even be a horse or dog, by operating a switch with his toes inside a shoe. If this ever becomes a common practice, the honorable art of rapid mental calculation will have indeed deteriorated to the level of a calculating-animal act or the acts of mediocre magicians who pose as psychics with awesome paranormal powers.

Postscript

The following letter from Steven Smith was published in *The New York Review of Books* (November 8, 1984) with my reply:

> I was very pleased with Martin Gardner's generous and perceptive review of my book, *The Great Mental Calculators,* but I hope that his extended discussion of deception in mental calculation will not lead readers to conclude that all the feats of professional calculators are trivial or fraudulent.

As Gardner remarks, some calculations are less difficult than they appear, such as extracting the roots of odd perfect powers (e.g., what is the twenty-third root of some 41-digit number where the root is known to be an integer). Such problems must be simpler than they seem, because they seem impossible.

In fact, these problems are trivial when the answer contains no more than two digits (as in the case above), but when the answer contains, say, eight digits, the difficulties are enormous. When Wim Klein extracts the thirteenth root of a hundred-digit number he must determine the logarithm of the first four digits of the power, divide it by thirteen and obtain the antilog. This gives him the first five digits of the root. (Klein has memorized the logs to five digits of the first 150 integers. The rest he gets by factoring, adding together the logs of the factors, and by extrapolation.) Then, to fix the last three digits of the root (in the case of an even number), he must divide the entire 100-digit number by 13, retaining only the remainder. His best time for this mental calculation is under two minutes. If you want to get a small idea of the difficulty involved, try dividing a hundred-digit number by 13 on paper as fast as you can and see how often you come up with the same answer.

Of course outright cheating occasionally occurs, but it is usually apparent to the knowledgeable observer. If someone appears to perform an evidently impossible calculation, they had best be able to give a credible account of how it can be done or I will start looking for tiny transmitters, concealed calculators, advance knowledge of the problem, peculiarities in the numbers, and so forth. I do hope that the chicanery of a few mountebanks will not cast a pall over the legitimate accomplishments of calculating prodigies, and that audiences will become more sophisticated so that they can distinguish between the impossible, the trivial, and the truly phenomenal.

There are also one or two points on which I disagree with Gardner, such as his contention that calculating prodigies in general employ a multiplication table of 100 by 100, but these can be left to the judgment of readers of my book.

Steven B. Smith

Soubès, France

Martin Gardner *replies:*

I certainly did not mean to suggest that the feats of the great mental calculators are trivial or fraudulent; indeed, I thought my review gave the opposite impression. In any case, I concur with everything in Smith's letter except his belief that most great calculators did not know the multiplication table to 100.

Consider Wim Klein and A. C. Aitken, two of the fastest mental calculators of recent times. Writing about them in his book *Faster*

Than Thought, B. V. Bowden said: "Both men have most remarkable memories—they know by heart the multiplication table up to 100×100, all squares up to $1,000 \times 1,000$, and an enormous number of odd facts, such as that $3,937 \times 127 = 499,999$."

Consider Arthur C. ("Marvelous") Griffith, a famous stage calculator. William Bryan and Ernest Lindley questioned him at length about his methods, reporting on them in their book *On the Psychology of Learning a Life Occupation.* Griffith, they write, "has [in his memory] a multiplication table complete to 130—and partial to almost 1,000 . . . is thoroughly acquainted with every prime and composite below 1,500, and can instantly give the factors of the latter."

Smith acknowledges in his book Klein's use of a 100 table, but thinks Klein a rare exception. A footnote recognizes Griffith's claim, but Smith doubts his honesty. On the other hand, Smith does not doubt any statement by a calculator who *denied* knowing the 100 table.

My contrary opinion rests on the extreme ease with which any mental calculator could memorize such a table, and the enormous aid it would be to him. Fred Barlow, in his book *Mental Prodigies,* quotes the French mathematician Edouard Lucas: "I formerly knew an instructor whose scholars, of eight to twelve years of age, for the most part knew the multiplication tables extended to 100 by 100 and who calculated rapidly in the head the product of two numbers of four figures, in making the multiplication by periods of two figures." In my opinion, it would be difficult for a great calculator to *avoid* memorizing the 100 table.

Smith is impressed by the fact that some calculators of the past, such as George Bidder, denied they knew such a table. I am more impressed by the notorious reluctance of professional calculators, like magicians and locksmiths, to give away all their trade secrets. Of course knowing a 100 table no more implies deception than knowing the 10 table.

25   Arthur C. Clarke

By almost anybody's standards, the world's two top writers who move effortlessly between science fiction and science fact are British-born Arthur Charles Clarke and his Russian-born American friend Isaac Asimov. "It is with great relief," Asimov once quipped (see *Science Digest's* March 1982 interview with Clarke), "that I add that he [Clarke] is three years older than I am and balder."

Since Clarke has written more than fifty books for the general public, it is good to have in *Ascent To Orbit* (Wiley, 1984) the cream of his scientific papers—articles that were too technical for his collections of popular essays.

As a bonus, Clarke has added sparkling commentaries, crammed with anecdotes and witty asides, from which one can extract the highlights of his distinguished career. After World War II, during which he was a radar instructor for the RAF, Clarke entered Kings College, London, graduating with honors in mathematics and physics. For thirty years he has made his home in Colombo, Sri Lanka, where he practices his favorite sport, skin diving, and is energetically at work on numerous projects.

Clarke's papers touch on all his major concerns: the excitement of space travel, the awesome possibility of contact with alien life forms, the technical wonders on the horizon, and, of course, the horrors of a nuclear holocaust. Clarke calls himself an optimist because he estimates humanity's chance of survival as 51 percent. In earlier books, he predicted communications with extraterrestrials by 2030 and the creation of artificial life by 2060. "The only way of discovering the limits of the possible," he wrote in *Profiles of the Future,* "is to venture a little way past them into the impossible."

This review appeared in *Science Digest,* June 1984, and is reprinted here, with changes, with permission.

Two other Clarke aphorisms:

"Any sufficiently advanced technology is indistinguishable from magic."

"Clarke's law: 'When a distinguished but elderly scientist states that something is possible, he is almost certainly right. When he states that something is impossible, he is very probably wrong.'"

Clarke has always taken delight in quoting from distinguished scientists who were careless enough to declare something impossible. In *Ascent to Orbit* he recalls for us England's Astronomer Royal, who declared that "space travel is utter bilge" shortly before the first Sputnik was launched. He remembers an early editor of *Amazing Stories* who said essentially the same in a balmy essay.

When the telephone was invented, the chief engineer of the British post office said: "The Americans have need of the telephone—but we do not. We have plenty of messenger boys." In contrast, Clarke adds, the mayor of an American city "thought that the telephone was a marvelous device and ventured a stunning prediction. 'I can see the time,' he said, 'when every city will have one.'"

Like his mentor, the British novelist and pioneer science-fiction writer H. G. Wells, Clarke has never hesitated to make prophecies, and his batting average is probably higher than Wells's. In a 1945 article in *Wireless World,* Clarke described with amazing accuracy how artificial satellites could be used for worldwide communications by placing them in a geosynchronous orbit about 42,000 kilometers (26,040 miles) above the equator. *Geosynchronous* means revolving in synchrony with the Earth's rotation, so that the space object remains fixed forever relative to a spot on the ground. Three such satellites, Clarke made clear, could provide global TV coverage, and the cost would be low because the satellite would draw energy from the sun. Almost twenty years went by before the first communications satellite was placed in a Clarke orbit.

Clarke is too modest in commenting on this famous *Wireless World* paper. He thinks it was just a lucky guess and that perhaps it advanced the cause of global communications by twenty minutes. Like prophet Wells, Clarke has his inevitable misses. It "seems unlikely," he wrote in the same article, that 20 years will elapse before atomic rockets are developed, capable of taking spaceships to the remoter planets of the solar system. But when Clarke is not writing fantasy, his misses are rare and minor. His 1949 book, *Interplanetary Travel,* the first in English to go into detail about modern spaceflight theory, swarms with astonishingly accurate guesses.

The book's specialized papers—on such topics as radar, electronics, TV wave forms, rockets, Lagrangian points, radio telescopes, and the dynamics of spaceflight—will be tough slogging for readers who know little about mathematics, but other papers are free of equations. A recent article on "The Space Elevator" or sky hook explains clearly how a geosynchronous space station could provide a simple, inexpensive way to launch spaceships.

"Help, I'm a Pentomino Addict!" is an amusing piece on how Clarke got hooked by a mathematical recreation called pentominoes. Clarke fans will recall that in his novel *Imperial Earth* a set of twelve tantalizing pentomino tiles becomes a symbol of nature's inexhaustible combinatorial possibilities. Less well-known is the fact that HAL, the ship's neurotic computer in *2001*, was originally filmed playing a pentomino board game. The game was actually sold by Parker Brothers, under the name Universe, with a scene from *2001* on the cover. At the last minute, the game in the film was changed to computer chess.

Now that Carl Sagan's research has intensified our awareness of the terrors that would follow nuclear war, Clarke's paper on "The Rocket and the Future of Warfare" is more timely today than when he wrote it more than thirty years ago. Our only hope, Clarke believes, echoing Wells, is rapid progress toward some sort of world community. It may take, Clarke has said elsewhere, an atomic disaster, with the loss of thousands of lives, to shock world leaders into sanity. His final paragraph says it all: "Upon us, the heirs to all the past and the trustees of a future which our folly can slay before its birth, lies a responsibility no other age has ever known. If we fail in our generation those who come after us may be too few to rebuild the world when the dust of the cities has descended and the radiation of the rocks has died away."

Did Sherlock Holmes Meet Father Brown?

26

Did Sherlock Holmes and Father Brown, England's two most famous crime solvers, ever meet? That they not only met but actually collaborated on a case was the startling conjecture of a paper read by Robert John Bayer at a 1947 meeting in Chicago of The Hounds of the Baskerville [sic].

Mr. Bayer, who lived in La Grange, Illinois, was the editor of a transportation magazine called *Traffic World*. His G. K. Chesterton collection (now owned by John Carroll University in Cleveland) was second in America only to that of John Bennett Shaw, the distinguished Sherlockian. Bayer's paper was printed for the Hounds as a chapbook of sixty copies. BSM reprinted it in its Winter 1981 issue, and now Magico has published an offset facsimile of the original.

Bayer argues with ingenious plausibility that when Father Brown told the story of "The Man with Two Beards" (in *The Secret of Father Brown,*) he deliberately concealed the fact that the private detective on this case was none other than Holmes himself. The detective is given a last name only—"Carver"—and we can assume that this was not his real name because he was on the scene incognito as a guest of Mr. Smith. Smith owned a bee farm in a town to which Father Brown gave the fictitious name of Chisham, and we are told that Carver was intensely interested in bees. The priest describes him as a "tall, erect figure with a long, rather cadaverous face, ending in a formidable chin." Could one ask for a better description of Holmes in his old age, after his retirement to beekeeping in Sussex? Although Bayer does not mention it, "Holmes" and "Carver" are both six-letter names with their vowels in the same places.

This review originally appeared in *Baker Street Miscellanea,* Winter 1984, and is reprinted here, with a postscript, with permission.

In "The Speckled Band," Holmes recalls a case involving Mrs. Farintosh and her opal tiara. A tiara owned by Mrs. Pulman is stolen in the Father Brown tale, and although its jewels are not identified, a woman who plays a role in the story is named Opal. Bayer suggests that Mrs. Farintosh and Mrs. Pulman are one and the same. Why would Father Brown alter the name? Because, Bayer reasons, the priest correctly solved the mystery by intuition whereas Carver's deductive solution was totally wrong. Out of great respect for the master, the kindly priest altered details to spare Holmes embarrassment.

In his introduction to Bayer's booklet, Vincent Starrett calls attention to a serious trouble spot in Bayer's thesis. When Holmes recalled the case of Mrs. Farintosh, he added that it took place before he and Watson met. But the Father Brown story speaks of motorcars, placing it at a much later date. I do not think this is hard to resolve. As Bayer speculates, Watson may have actually written up the case only to have his agent, Conan Doyle, set it aside. Why? Opal was a devotee of Spiritualism—which Father Brown called "nonsense." As England's fugelman for Spiritualism, Conan Doyle would have had a motive for suppressing Watson's story. Can we not go a step further? Although the case occurred late in Holmes's life, Conan Doyle could have slyly inserted a reference to it in an earlier story, falsifying the date so readers would never connect Mrs. Farintosh with Mrs. Pulman. Sherlockians have long been suspicious of this passage. If the Farintosh case occurred before Watson's time, how did Helen Stoner, who came to see Holmes, get his Baker Street address from her friend Mrs. Farintosh?

There is another discrepancy in Bayer's thesis that Starrett failed to notice. Bayer reports Father Brown's remark that Carver had "bright" eyes, but he does not supply the full sentence: "The brow was rather bald, and the eyes bright and blue." As all Sherlockians know, Holmes's eyes were the same color as Father Brown's—gray. Here again, I think we may safely assume that the priest, anxious to spare Holmes humiliation, would have altered the color of Carver's eyes to hide his identity.

Let me close with a speculation of my own. In the first paragraph of the Father Brown story "The Eye of Apollo," the book's first edition refers to the priest as "Reverend J. Brown." For reasons unknown, the J was removed from all later editions. Is there a J. Brown in the canon? Yes indeed! In "The Adventure of the Six Napoleons," we learn that a Joshua Brown, of Chiswick, purchased one of the six plaster busts.

Is it possible that a young Father Brown, perhaps not yet a

man of the cloth, knowing that one of the busts contained the black pearl of the Borgias, was working independently on this case? You may recall that Joshua Brown cooperated with Holmes by locking the doors of his house to await the arrival of a thief. Did Watson conceal the fact that Joshua Brown was an amateur detective, who later became the famous priest? Like Father Brown in later years, was Watson protecting a man's reputation—keeping from his readers the fact that Father Brown had once tried unsuccessfully to solve a crime that called for an older detective of greater experience? Surely the possibility calls for careful investigation.

Postscript

Sherlock Holmes was a philosophical theist, and although he never attended a church he always treated the Roman Catholic faith with respect. Is it possible, Sherlockian Stefan Kanfer has recently asked, that Father Brown may have been partly responsible for this attitude?

In *The Hound of the Baskervilles* (Chap. 2) we learn that Holmes had been "exceedingly preoccupied by that little affair of the Vatican cameos," during which he was extremely anxious "to oblige the Pope." In "The Adventure of Black Peter" Watson tells us that 1895 was the year of Holmes's "famous investigation of the sudden death of Cardinal Tosca—an inquiry which was carried out by him at the express desire of His Holiness the Pope." In "The Final Problem" Holmes actually disguises himself as an aged Italian Catholic priest to protect himself against an attack by Professor Moriarty.

Sam Brown, a Scotland Yard inspector, plays a role in *The Sign of Four*. Could he have been Father Brown's brother—one whose profession stimulated the priest's early interest in crime? The question suggests how much the Father Brown stories lend themselves to the same kind of research as Watson's chronicles. For a first attempt at such exegesis see my *Annotated Innocence of Father Brown* (Oxford University Press, 1987).

27

Richard Feynman

"The stories Feynman tells about himself would make a book," an interviewer wrote in 1963. Here, twenty-two years later, is that book. It is not an autobiography. You will learn nothing from it about the work on quantum mechanics that earned Richard Feynman a Nobel prize in 1965. What you will learn is a great deal about the flamboyant personality of a great theoretical physicist: his amazing range of interests, his unusual hobbies, his enthusiasms and animadversions, his dislike of pompous fools, his unpredictable behavior and, above all, his fondness for outrageous comedy.

Feynman's famous three-volume *Lectures on Physics* and his marvelous little book *The Character of Physical Law* were based on taped lectures. *Surely You're Joking, Mr. Feynman!* (Norton, 1985) is a taping of Feynman's uninhibited conversations over the years with his friend Ralph Leighton. "That one person could have so many wonderfully crazy things happen to him in one lifetime is sometimes hard to believe," Leighton writes in his preface.

The reason they happened is that Feynman has a knack of creating his own adventures. When, for instance, he was a young man at Los Alamos, working on the top-secret bomb, he discovered a hole in the perimeter fence. Anyone else would have reported this to the authorities and that would have been the end of it, but Feynman is incapable of passing up a chance for a lark. Out of the main gate he goes, back through the hole, then out again. He keeps this up until the sergeant at the gate suddenly realizes that some strange character is always going out but never coming in.

Feynman's expert ability to pick locks followed naturally from his lifelong passion for puzzles. To demonstrate how loose security

This review originally appeared in *Nature*, 25 April 1985, and is reprinted here with permission.

This review originally appeared in *Nature*, 25 April 1985, and is reprinted here with permission.

was at Los Alamos, he would secretly open the safes of top scientists, leaving little notes that said such things as "I borrowed document LA4312—Feynman the safecracker." But such idiosyncracies were not appreciated by everyone, least of all the military. When three psychiatrists tried to test Feynman for Army service they quickly found themselves being tested. "How much do you value life?" one of them asked. "Sixty-four," Feynman answered. The Army's final verdict: mentally deficient. (The chapter on this banter, headed "Uncle Sam Doesn't Want *You!*" is one of the book's funniest.)

My favorite anecdote, however—it is vintage Feynman—concerns a joke his friends once tried to play on him. In Japan Feynman had learned some Japanese, and in Portugal he gave lectures in Portuguese. At a party someone thought it would be amusing to see how this "man of a thousand tongues" would react if a Caucasian lady, who grew up in China, greeted him in Chinese. "Ai, choong, ngong jia!" she said with a bow. Taken aback, Feynman swiftly decided the best thing to do was imitate the sounds she made. "Ah, ching, jong jien!" he replied, returning the bow. "Oh, my God!" the lady exclaimed. "I knew this would happen. I speak Mandarin and he speaks Cantonese!".

From childhood on, anything mysterious or puzzling instantly aroused Feynman's curiosity. Growing up in Long Island, New York, he taught himself how radios work and he became the neighborhood's youngest radio repairman. He invented ingenious experiments with ants to work out how their brains were programmed. Intrigued by hypnotism, he allowed himself to be hypnotized. To experience out-of-body hallucinations, he floated in a sensory-deprivation tank. Always, wherever he was or whatever he did, the wheels in Feynman's brain never stopped whirring. Every experience posed new challenging questions. What goes on here? Why does this work? Can it be done better? The best example echoes Newton and the apple (no, that story is not a myth). Feynman once watched someone toss a plate in the air and noticed that its wobble went around faster than the plate. It started a train of thought that eventually led to the Feynman diagrams for particle interactions and to the work that won the Nobel award.

The book swarms with delightful glimpses of the famous: Einstein, Bohr, Wheeler, Wigner, Oppenheimer, Teller, Pauli, Compton, Fermi, Gell-Mann, and many others. (It is said that the California Institute of Technology, where Feynman has taught since 1950, hired Murray Gell-Mann so Feynman would have someone to talk to.) And it was, apparently, a casual remark of one of them, John von Neumann, that gave Feynman a sense of political detachment that he claims has kept him happy ever since. You do not have to feel responsible, the great mathematician told him, for the state of the world. Not for its social and political insanities, perhaps, but trying to understand how nature works on her deepest level is a responsibility—perhaps pleasure is a better word for what Feynman feels—that he surely has not taken lightly.

28

O amazement of things—even the least particle!
—Walt Whitman, "Song at Sunset"

Theoretical physicists are in a state of high excitement these days—
and for good reason. New discoveries in particle physics, combined
with brilliant theoretical invention, suggest that they are on the
verge of nothing less than explaining everything.

Well, not exactly Everything—but everything possible for
physics to explain. More precisely, they believe they are close to
constructing a unified field theory that will describe exactly how
the universe, almost instantly after the Big Bang, acquired all the
particles and forces that allowed it, some fifteen billion years later,
to grow into the universe we know. A few adventuresome theorists
think they may soon be able to explain how the primeval explosion
itself was caused by a random quantum fluctuation of Nothing.

One of the two books under review is *Perfect Symmetry* (Simon and Schuster, 1985), by the American physicist Heinz Pagels,
whose previous book, *The Cosmic Code,* is one of the best introductions to quantum mechanics I have ever read. The other volume is
Superforce (Simon and Schuster, 1984), by the British physicist
Paul Davies, author of many earlier books that are models of science
writing for laymen. Both books are admirable up-to-the-minute accounts of the search for what Pagels calls the "Holy Grail." Those
who work in "the shadowy world of fundamental physics," Davies
writes in his first paragraph, are about to complete their long quest
"for a prize of unimaginable value—nothing less than the key to the
universe."

Is it really true that physics may be nearing the end of a road,
"going for broke," as Pagels puts it? Of course, there will remain the
infinite problems on what Pagels calls "the frontier of complexity"—such trifles as explaining how the basic forces and particles

manage to get together and write books about themselves—but there may be nothing more to learn on a bedrock level. The situation will be something like that of plane geometry. All its theorems are implied by its axioms, but the number of theorems yet unknown is infinite. A unified field theory would in no way be the end of science or technology. There will be endless inventions to make, endless worlds out there in space to explore. It would only mean the end of the search for fundamental laws.

Unfortunately, physicists have been in previous states of euphoria about reaching the end of the road. In the late twenties everything seemed wonderfully simple. Maxwell's field equations had explained electromagnetism. Einstein's field equations had explained gravity. And Einstein was hard at work on a theory to unify these two forces. All matter was made of atoms that contained just two kinds of particles: protons in the nucleus and electrons whirling around in paths described by probability waves. "Physics as we know it will be over in six months," said Max Born, one of the great architects of quantum mechanics.

In 1932 nature began to look shaggy again. A new particle, the neutron, was found hiding in the nucleus. Paul Dirac's theoretical work implied that the electron had an antiparticle twin exactly like it but with a positive charge, and, sure enough, in 1932 the positron was found. Then came the deluge. Dozens of entirely unsuspected particles began to turn up as physicists started clashing particles together at high speeds in the new accelerators—particles that had, in Robert Oppenheimer's words, an "insulting lack of meaning." When it appeared later that protons and neutrons were made of smaller particles, the physicist Murray Gell-Mann, drawing on *Finnegans Wake,* called these "quarks" ("three quarks for Muster Mark!").

Two entirely new forces arrived on the scene. Neutrons and protons were found to be held together in the nucleus by a "strong force," much stronger than gravity or electromagnetism but operating at extremely small distances. In radioactive decay, when a neutron decays into an electron, proton, and an antineutrino, the process was found to be controlled by a "weak force."

Slowly, over the next four decades, the hundreds of newly discovered particles began to make sense. It now appears that everything material is made from combinations of six kinds of quarks and six particles called leptons. The four forces, or "interactions," as physicists prefer to say—gravity, electromagnetism, and the "strong" and the "weak" forces—are transmitted by a third set of more ghostly particles called bosons.

The leptons are pointlike particles that are influenced by the weak force. For example, when radioactivity takes place, and a neutron decays into a proton, the weak force creates two leptons, an electron and an antineutrino. Leptons have no known interior structure, but they have mass and a curious quantum property called spin, to which I will soon return. The electron is the most important lepton because it is part of all atoms and because it carries electrical charge. (Electric currents are produced by moving electrons.) The muon is the funniest lepton. It has been called a fat electron because it is just like an electron in every respect except that it weighs more than two hundred times as much. "Consider the muon," Columbia University's Isidor I. Rabi once began a lecture. "Who ordered *that?*" The tauon, discovered in 1976, is even odder. It is a fat muon more than 3,500 times as massive as the electron. When the three particles are involved in certain interactions, each is associated with its own kind of neutrino.

Neutrinos, which make up the other three leptons, are as close to nothing as a particle can get. They have no electric charge and no known mass. Their only property seems to be spin. Quantum spin is something like the spin of a top but much more mysterious. (How can a point spin?) All three neutrinos are left-handed in the sense that when they move away from you their spin is counterclockwise. Each of the six leptons has a twin antiparticle of opposite spin (and opposite charge if not a neutrino); so if antiparticles are considered, there are twelve leptons.

The six quarks come in six types or "flavors": up, down, charmed, strange, top (or truth), and bottom (or beauty). These words have no connection with their ordinary meanings. They are simply colorful terms for properties of quarks that can only be described by mathematical expressions. All quarks are believed to be pointlike, though some theorists have speculated that they are made of smaller particles, and a few think there may even be an infinity of sublevels. Each quark has mass, spin, a fractional electric charge of plus two-thirds or minus one-third, and a different kind of charge called color. It differs from an electric charge in that it applies to a new property of matter that comes in three varieties: red, green, and blue. Of course the quarks do not have colors in the usual sense, but the color names are useful because of analogies between the mixing of quark colors and the mixing of ordinary colors.

Each quark has its antiquark of opposite charges and spin. The heavier a particle, the harder it is to produce in an accelerator, and for this reason the heaviest quark, the top quark, was the last to be detected. Carlo Rubbia, heading a team of scientists at CERN (The European Laboratory for Particle Physics, in Geneva) announced evidence for the top quark late in 1984.

Quarks join together in only two ways, pairs or triplets, to make composite particles called hadrons. Hadrons are influenced only by the strong force. It binds together the quarks inside them and also binds them to one another. Quark doublets (each consisting of a quark and an antiquark) form hadrons called mesons. The triplets (three quarks) are the baryons, of which the most important are the proton (two up quarks and one down) and the neutron (two down and one up). Every possible combination of doublets and triplets is a hadron that has been experimentally verified.

Particles that carry the four forces go by various names: bosons, exchange particles, interaction particles, carrier particles, virtual particles, and ghost particles. They can be thought of as particles that zip rapidly back and forth between the "real" particles. The photon carries the electromagnetic force, the graviton (not yet detected) carries gravity. The weak force has three carriers: a positively charged W particle, a negatively charged W particle, and a chargeless Z particle. A 1984 Nobel prize went to Rubbia and Simon van der Meer for their observations at CERN of all three particles. The strong force that glues together the quarks inside the hadrons—and the hadrons to each other—is carried by eight kinds of bosons appropriately called gluons. (Pagels uses the term *gluon* for all exchange particles and *color gluons* for carriers of the strong or color force.)

Every force is described by a field such as the magnetic field around a magnet or the gravity field around the earth. Every field has its carrier particle, and every particle, carrier or otherwise, has its field. Pagels's chapter on fields is especially useful in making clear the overriding importance of fields and their symmetries.

Symmetry is that property of a structure which remains the same if you perform a certain operation on it. For example, the letter H has 180 degree rotational symmetry because if you turn it upside down it does not change. It also has left-right symmetry because it is unaltered by mirror reflection. It has glide symmetry because it stays the same if you slide it along the page. The letter F has glide symmetry but lacks rotational and reflection symmetry.

Fields share all the symmetry properties of their particles. A particle is simply a property of a field. Newton's atoms were like hard little billiard balls, but in quantum mechanics such atoms have totally dematerialized. For a crude analogy, think of a sheet of paper filled with hundreds of parallel creases. Imagine that another set of parallel creases, at right angles to the first set, glides across the page. At spots where the creases intersect, points move across the page. Particles resemble those points. They are created by the movements of fields. "The world according to this view," Pagels writes, "is a vast arena of interacting fields manifested as quantum particles flying about and interacting with each other."

In our analogy the fields are made of paper. What are quantum fields made of? The question is meaningless. They are not made of anything. They are irreducible in the sense that they cannot be reduced to something more fundamental. They are pure mathematical concepts. They just are.

The first great modern unification theory was James Clerk Maxwell's joining of magnetism and electricity. They were thought to be independent forces until Maxwell's field equations (field equations describe how a field changes in space and time) combined them. Gravity and inertia were similarly considered different until Einstein showed them to be manifestations of a single field. Einstein spent the last part of his life vainly trying to unify gravity and electromagnetism. Nobody can blame him for failing because the data he had at hand were too scant.

In 1961 Harvard's Sheldon Glashow laid the groundwork for a unification of electromagnetism and the weak force. The theory was completed by Steven Weinberg and Abdus Salam, working independently, and for this the three received a 1979 Nobel award. The new field is called the electroweak field. The theory predicted the W and Z particles, which carry the weak force. Their discovery was such strong confirmation of the theory that it is now accepted as standard particle physics. Formerly there were four basic forces. Now there are three—gravity, the strong force, and the electroweak force.

The next step is to unify the electroweak force with the strong. Hundreds of attempts to do this, known as GUTs (grand unified-field theories) are now being proposed. The simplest and most promising is one constructed by Glashow and Howard Georgi in 1973. It makes several predictions, none as yet verified.

The final step, of course, would be a field that unified all the forces, including gravity. Such a theory would be extraordinarily difficult to confirm because it would require accelerators more powerful than any now conceivable, but that has not inhibited the theorists, especially the younger ones. They are enthusiastically at work on what are called either super-GUT theories or supersymmetry theories, SUSYs for short. It may turn out that the only arena in which confirmations can be found is in interstellar space, where stars and black holes produce extreme temperatures and energies beyond the power of earthly instruments. Pagels recalls a motto he saw on a student's T-shirt: "Cosmology takes GUTs."

To understand how cosmology is involved it is necessary to grasp the concept of breaking symmetry. Weinberg likes to explain it with a marble balanced on top of the glass mound at the base of certain bottles. The structure has circular symmetry in the sense that bottle and marble look the same from all sides. But the structure is unstable. The marble rolls off the mound to one side, breaking the original symmetry. Abdus Salam likes to explain it with a group of people sitting symmetrically around a circular table. Infront of each is a dinner plate, and between every adjacent pair of dinner plates is a salad plate. The situation is symmetrical until the hostess decides whether to reach right or left for the salad. As soon as she decides, everybody reaches the same way. Left-right symmetry is broken. In cosmology, the symmetries that break are much more complicated. They are properties of equations that cannot be expressed in visual pictures.

SUSYs assume that immediately after the Big Bang, when the temperature of the universe was unthinkably high, elegant symmetries prevailed, then were broken as the universe rapidly expanded and cooled. Let us run this script backward in time. When temperature rises to a certain point, electromagnetism and the weak force become one force. Go farther back in time, when temperatures are still higher, and the electroweak force fuses with the strong. Go back some more, to less than a nanosecond (one-billionth of a second) after the Big Bang. All forces are now a single force field, perhaps with a single superparticle. This is the force that provides the title "superforce" of Davies's book and the "perfect symmetry" title of Pagels's book.

Water has high rotational symmetry because no matter how you turn it it looks the same. But when water freezes into a snowflake a "phase transition" occurs. It loses its rotational symmetry to acquire a beautiful hexagonal pattern that now must be turned in 60-degree increments to make it look the same. Broken symmetries like this, though far more complicated and on a vaster scale, are believed to have occurred while the universe cooled. "Our universe today," Pagels writes, "is the frozen, asymmetric remnant of its earliest hot state."

My favorite model of symmetry breaking is an old stunt with playing cards. On a tablecloth, using great care and patience, it is possible to balance four cards on their long edges so they radiate out from a point like the arms of a cross. Gently push four more cards into the gaps to make a wheel with eight spokes. Add more cards one at a time. The more you add the more stable the structure should become until finally you have a wheel with fifty-two spokes. It is not easy to form. It helps if you give the deck a slight bend. You can anchor the outer ends of the first four cards with small objects such as checkers or chessmen and later remove them. The trick is prettier if you place the cards so they all face around the circle in the same direction.

Big bang your fist on the table. The jar will break the symmetry, collapsing the structure into a lovely rosette that is either right- or left-handed. Many physicists believe that an event similar to this explains why the universe we know is made of matter. Originally the cosmos was symmetrical with respect to matter and antimatter (matter made of antiparticles). When the symmetry broke, the universe collapsed into matter, but it could just as easily have gone the other way.

Davies devotes a colorful chapter to a popular SUSY known as a generalized Kaluza-Klein (KK) theory. When I was writing my *Ambidextrous Universe* in the early sixties I came across a forgotten theory proposed in the twenties by Polish physicist Theodor Kaluza and the Swedish physicist Oscar Klein. They tried to unify gravity and electromagnetism by assuming a fourth spatial dimension, closed like a circle and with a radius smaller than an atom's. They claimed that electromagnetism is actually a form of gravity, its waves moving in this unseen dimension of space. Think of every point in space as attached to an incredibly tiny circle that goes in a direction impossible to visualize. As Davies describes the theory, "what we normally think of as a point in three-dimensional space is in reality a tiny circle going round the fourth space dimension." From every point in space, as Davies puts it, "a little loop goes off in a direction that is not up, down, or sideways, or anywhere else in the space of our senses. The reason we haven't noticed all these loops is because they are incredibly small in circumference." If time is added as a coordinate, the little loop becomes a threadlike cylinder in five-dimensional space-time. Gravity waves spiral around these threads in helixes that have either of two mirror image forms. One direction produces positive charge, the other a negative charge.

I discussed the KK theory not only because I found it whimsical but because it explained positive and negative charge according to left-right handedness and also gave a reason why charge comes in discrete units. (Something going around a loop has to go completely around it to get back where it started.) Einstein took the theory seriously. "The idea of achieving [a unified theory] by means of a five-dimensional cylinder world," Pagels quotes him as writing to Kaluza, "never dawned on me. . . . At first glance I like your idea enormously." Eventually Einstein decided the theory was wrong. In 1963 I asked several top physicists what they thought of the KK theory. None had even heard of it.

You can imagine my surprise fifteen years later when the theorists suddenly remembered KK. It turns out that a simple, beautiful way to explain the properties of particles is to generalize KK by turning the tiny circles into seven-dimensional spheres! If this conjecture proves fruitful, all interactions and particles become aspects of one superforce shimmering around in a space-time of eleven dimensions. Three are the ones we know, seven are the "compacted" dimensions of the invisible hyperspheres, and the eleventh is time.

It is good to realize that when physicists talk about spaces they do not always mean spaces that are physically real. They are usually artificial spaces devised to simplify calculations. No physicist thinks that the curves he uses to graph functions on two-dimensional paper are "out there" in physical space or that the probability waves of quantum mechanics (they are waves in imaginary "phase spaces" of high dimensions) are out there like water or sound waves. Probability waves exist only in the minds and discourse of physicists. On the other hand, the higher spaces of the KK theories (some have more than seven new dimensions) could be as real as our familiar space of three dimensions. On this dark question KK enthusiasts are sharply divided. No one can even think of an empirical test that might settle the matter.

It is dangerous, Sherlock Holmes once said to Dr. Watson, to theorize without adequate facts. Nevertheless, it is essential to science that dangerous theorizing constantly go on, and Davies is among those superoptimists who think the ultimate unification may be as imminent as Billy Graham thinks the Second Coming is. Pagels, too, is optimistic, though more cautious:

A whole community of very smart scientists may have talked themselves into a theory of the very early universe that in the future (with the wisdom of hindsight) will be seen as a fantasy based on incomplete information and imaginative extrapolation. Theory building, while it creates a framework for thought, is never a substitute for experiment and observation. The new high-energy accelerators and telescopes currently on the drawing boards will tell us a lot about whether or not these ideas are correct.

Sometimes I wish that this book about the current ideas of physics and cosmology could be published like a loose-leaf notebook. That way, pages could be discarded and replaced with new pages describing better ideas when they come along. Much of our current scientific thinking about microscopic physics, the "wild ideas" and cosmology is probably wrong and will have to be discarded. Maybe in the future there will be a major revolution in physics that will revise our whole idea of reality. We may look back on our current attempts to understand the origin of the universe as hopelessly inadequate, like the attempts of medieval philosophers trying to understand the solar system before the revelations of Copernicus, Kepler, Galileo and Newton. What we now regard as "the origin of the universe" may be the temporal threshold of worlds beyond our imagining. But it is also possible that we are near the end of our search. No one knows.

The wildest of all the wild ideas now being tossed about is based on the assumption that a vacuum is not pure nothing because it is saturated with quantum fields. Even if we think of the vacuum as spaceless and timeless, there may be some sort of mathematical structure to it, more fundamental than space and time, with its associated quantum laws. Given enough time—whatever that means!—there is some degree of probability that the structured "nothing" will become unstable. A single spot of "something," produced by a random quantum event, will explode into a space-time universe. Many physicists, Pagels among them, are playing with such notions, while others think it is a waste of time—as absurd as looking for something north of the North Pole.

This view of the universe as a "free lunch" is close to the medieval doctrine of creation ex nihilo, though we must be careful now to distinguish between points of view hotly debated by the Scholastics. There were many subtle variations, but the essential conflict was between those who argued that God created the universe from absolutely nothing and those who argued that God made use of a formless, eternally existing primal matter. The two points of view are mirrored today in the views of cosmologists who are unwilling to consider creation by an Outsider. There may exist from all eternity a primal Mother Field of space and time, capable of giving rise to a singularity—the random quantum event previously mentioned—that explodes into a universe. Or there may be no field at all—just nothing, completely empty of space and time. Somehow behind this nothing are quantum laws that can spontaneously produce a space-time field that in turn explodes into the universe.

Observe that nothing is not absolutely nothing for either the Scholastics or a secular cosmologist. All the theologians of the Middle Ages assumed an eternally existing Creator. Today's cosmologists must assume, at the least, eternally existing quantum laws.

"[The] unthinkable void," Pagels writes, "converts itself into the plenum of existence—a necessary consequence of physical laws. Where are these laws written into that void? What 'tells' the void that it is pregnant with a possible universe? It would seem that even the void is subject to law, a logic that exists prior to space and time."

It would seem indeed! Because the existence of laws is not nothing, the new physics adds nothing to help answer the unanswerable superultimate question: Why is there something rather than nothing?

Davies also struggles with this impenetrable mystery, ending his book on an enigmatic note of pantheism. Although science may in a sense explain the universe, "we still have to explain science. The laws which enable the universe to come into being spontaneously seem themselves to be the product of exceedingly ingenious design. If physics is the product of design, the universe must have a purpose, and the evidence of modern physics suggests strongly to me that the purpose includes us."

Perhaps. But I cannot see how anything in modern science suggests this more than does anything in ancient science. That the universe displays an incredible order, not made by us, was as obvious to a Roman atheist such as Lucretius as it is to a modern theist. The only difference is that today cosmic evolution has pushed speculation about the source of this order down to the quantum level and back to a primordial fireball. For all science and reason can tell us, a mindless Mother Field may have generated precisely the patterns we find and may be as indifferent to human destiny as it is to the fate of a symmetrical snowflake.

On this question, the deepest in philosophy, Pagels does not tell us his private views. Nowhere in his splendid book does he consider the possibility of a Mind outside the cosmos, although he confesses that the universe continues to haunt him. "This sense of the unfathomable beautiful ocean of existence drew me into science. I am awed by the universe, puzzled by it and sometimes angry at a natural order that brings such pain and suffering. Yet any emotion or feeling I have toward the cosmos seems to be reciprocated by neither benevolence nor hostility but just by silence. The universe appears to be a perfectly neutral screen onto which I can project any passion or attitude, and it supports them all."

Yet Pagels's epigraph for his book is that spine-tingling first verse of an old religious document from the Middle East that tells how the world was once without form and void, how darkness was on the face of the deep, and how the spirit of God moved over the waters. The superultimate question remains as stark as ever. Back of the Mother Field, behind space and time and the laws of quantum mechanics, is there something analogous to human consciousness? Or is the universe, as G. K. Chesterton once wrote, "the most exquisite masterpiece ever constructed by nobody"?

"Someday," Pagels writes, "(and that day is not yet here) the physical origin and the dynamics of the entire universe will be as well understood as we now understand the stars. The existence of the universe will hold no more mystery for those who choose to understand it than the existence of the sun."

I can (with effort) buy the first sentence but not the last. Theorems of geometry are not very mysterious. It is a formal system's axioms that pop like magic out of nowhere. A set of laws with the awesome power to blast into reality a cosmos containing life forms as fantastic as you and I is to my mind so staggering a vision that it makes the origin and dynamics of a star seem as trivial as the origin and dynamics of an eggbeater.

Postscript on Superstrings

When I wrote this review the currently fashionable KK theory required seven extra space dimensions. Since then, superstring theory has been formulated in which the number of extra, curled-up dimensions drops to six.

On the last page of my *Ambidextrous Universe* (Basic Books, 1964) I quoted a famous remark by Niels Bohr. The German physicist Wolfgang Pauli had finished lecturing in 1958 on a new conjecture about particles. Bohr arose and said: "We are all agreed your theory is crazy. The question which divides us is whether it is crazy enough to have a chance of being correct. My own feeling is that it is not crazy enough."

Well, the crazy theory is here, and I suspect that if Bohr were around he would be entranced. Edward Witten, a noted Princeton physicist, has described superstring theory as "beautiful, wonderful, majestic—and strange." He thinks the next fifty years will be devoted to work on the theory's implications and possible testing. On the other hand, Sheldon Glashow, who shared a Nobel prize with Steven Weinberg and Abdus Salam for unifying the weak and electromagnetic forces, has jingled:

> Please heed our advice
> That you too are not smitten.

> The book is not finished,
> The last word is not Witten.

To suggest how crazy superstring theory is, let me give a hopelessly inadequate precis of its major ideas. All matter is believed to be made of pointlike particles that belong to two classes: the quarks (of which there are six varieties, not counting their antiparticles) and the leptons (also six varieties, not counting antiparticles). The most important lepton is the electron. Until now it has been regarded as a geometrical point with no known spatial structure— only quantum properties.

In superstring theory, basic particles are modeled not as points but as inconceivably tiny one-dimensional strings. In the most promising superstring theory they are closed like rubber bands. These loops should not be thought of as made of smaller entities, the way elastic bands are made of molecules. They are the quantized aspects of string fields. The infinitesimal loops move, rotate, and vibrate in a space of ten dimensions: one of time, three the familiar dimensions of our experience, and six that are "compacted" in the sense that they are curled into invisible hyperspheres at every point in three-dimensional space. When the loops move they follow geodesics that trace minimal surface areas on what is called a "world sheet."

Superstring theory is the latest dramatic instance of how mathematicians construct theorems and formal systems of no known utility—theorems and systems that then suddenly turn out to have practical applications. Notable past examples include the Greek conic-section curves; Riemann's work on non-Euclidian spaces, work that became so essential in relativity theory; work on matrices, group theory, and statistics that became part of quantum mechanics; and Boolean algebra, which underpins the designing of computer circuitry. In superstring theory it is the work of topologists on two-dimensional surfaces embedded in higher-dimensional spaces. To their vast surprise, topologists now find themselves frantically teaching topology to particle physicists while simultaneously struggling to master quantum mechanics. Weinberg recently speculated that some mathematicians sell their souls to Satan in exchange for information on what new areas of pure mathematics will have profound applications in science!

Are the higher dimensions of superstring theory "real," or are they artificial constructs like the infinitely dimensional Hilbert spaces of quantum mechanics? Physicists are dividing over this question. Some see the compacted dimensions as no more than useful artifacts. Others see them as no less real than three-dimensional space.

A similar debate concerns the reality of superstrings. At this moment, any prediction about the outcome would be foolhardy. Ernst Mach, the German physicist who so strongly influenced Einstein, could not believe that atoms and molecules were anything but mathematical abstractions—useful, yes, but no more "out there" than the curves that represent a functional relationship between two variables are out there. Atoms now have passed from theoretical entities to "observables" that can be seen in microscopes. Superstrings obviously are not like ordinary strings or elastic bands; nevertheless, they could model structures as much "out there" as molecules, trees, and stars.

String theory goes back to the late 1960s, when no one took it seriously. It was not until about 1980 that John Schwartz of Caltech, and Michael Green of London's Queen Mary College, transformed strings into superstrings by combining them with GUTs based on the supersymmetry of force fields immediately after the Big Bang and on the symmetry breaking that occurred as the superhot universe expanded and cooled. A few years later Green and Schwartz succeeded in purging superstring theory of numerous inconsistencies that had plagued it. It was this purging that ignited the current big bang of interest in superstrings.

For the first time, apparently, there is now a plausible, elegant way to account for all the forces of nature as well as for the properties of all the particles, especially the as yet undetected graviton. The graviton belongs to a family of bosons, the so-called virtual particles that are the carriers of forces. Other GUTs incorporate gravity, but superstring theory is the first to require gravity as an essential part of the theory. The graviton is the simplest mode of vibration the little loops can have. Without gravity, the strings fall to the ground.

Superstring theory is what some physicists like to call a TOE, an acronym for Theory of Everything. Of course it doesn't really explain everything. For one thing it has not explained (yet) why the universe, after the primeval fireball, curled up six space dimensions into tight little hyperspheres while the other three, along with time, expanded. And of course it does not explain why nature selected equations that describe the behavior of shimmering strings to build a universe, including you and me. Why is there something rather than nothing? And why is that something mathematically structured the way it is? These are metaphysical questions that clearly, at least to me, are in principle beyond the reach of both science and philosophy.

My own opinion is that the Big Bang was a laboratory experiment, that TOE refers to the big toe of a hyperphysicist who used her toe to press the button.

W. V. Quine

29

Not many philosophers attempt autobiographies—Bertrand Russell and George Santayana are the most notable modern exceptions—so it was a rare event on June 25, the 77th birthday of Willard Van Orman Quine, when *The Time of My Life* (MIT Press, 1985) was published. Though little known outside academia, Quine is the most distinguished American-born philosopher since John Dewey. His views have been enormously influential and, to this day, continue to generate heated and fruitful controversy.

In earlier ages, philosophers were expected to have deep opinions on almost everything, but now they are as specialized as scientists. Quine's specialties are set theory, logic, semantics, and linguistics. Like his friend and mentor, Rudolf Carnap, the most famous of the Vienna Circle empiricists, Quine has no wisdom to impart about aesthetics, ethics, political philosophy, or religion. For this reason, his autobiography has less in common with those of generalists such as Russell and Santayana than with the autobiography of an opthalmologist or a chess grand master.

Even with respect to his special interests, Quine seldom has much to say, except for one brief chapter on his "Mathematical Logic," or "ML," as it is known. This is a formal system similar to "Principia Mathematica," or "PM," constructed by Russell and Alfred North Whitehead, but simpler and more elegant. Quine intended it to avoid the paradoxes that marred PM, but, unfortunately, ML proved to have paradoxes of its own. Quine partly banished them from the book's revised edition and corrected little mistakes such as a careless reference to Paul as one of the apostles. Hao Wang, Quine's brilliant pupil, finally completed the repairs.

Quine was born in Akron, Ohio, in 1908, and it was while he

This review originally appeared in *The Boston Globe,* 7 July 1985. © 1985 by Martin Gardner. All rights reserved.

was a mathematics major at Oberlin College that the three volumes of PM dropped on his head like a thunderbolt. Scholarships at Harvard allowed him to complete a doctorate under Whitehead himself. "That's ripping, old fellow. Right jolly!" said Whitehead when Quine explained his choice of topic. "Bertie [Russell] thinks I'm muddle-headed," Whitehead told him on another occasion, "but I think Bertie's simpleminded."

Eventually, Quine became chairman of Harvard's philosophy department, and it would take paragraphs to list all the academic honors he has received since. During World War II, he enlisted in the Navy, serving several years in Washington as a cryptanalyst. A WAVE who worked in his office became his second wife after a bitter divorce that Quine covers in a grim chapter titled "Sturm and Drang."

As a boy, Quine loved to draw maps, and his interest in geography combined naturally with philately. For a while, he and a friend published a little periodical called *O.K. Stamp News.* In adult life, his passion for collecting stamps became, as he puts it, a passion for collecting countries. There are long stretches in Quine's book, as he crosses more than one hundred national borders, that read like the trivial details of a travel diary.

Quine sees himself as a man preoccupied with the precision, beauty, and simplicity of formal logic, as taciturn, easily bored, introverted, but with "little talent for soul searching." He has always lived frugally and simply, and, although he dislikes personal confrontations, he recognizes that much of his reputation has been aided by the controversies he has initiated.

Quine's most notorious argument was with Carnap over what logicians call the analytic-synthetic distinction. Almost all philosophers since Hume have contrasted analytic sentences that are true in virtue of the meanings assigned to their words ("All black cows are black") with synthetic sentences that require observation of the world before they can be deemed true ("Some cows are brown"). No one denies that the distinction has a fuzzy dividing line, like the line between night and day or between spoons and forks, but Quine obviously means more than that. His subtle attacks on analyticity have even given rise to the use of his name as a verb. "To Quine," it is said, means "to repudiate a clear distinction." Mathematician John Kemeny, the former president of Dartmouth, once described Quine's efforts to undermine Carnap on this point as "the most important losing battle in the history of modern philosophy." Entire books have been devoted to this battle. Both sides are defended at length in *The Philosophy of W. V. Quine,* edited by Lewis Edwin

Hahn and Paul Arthur Schilpp, volume 18 in the distinguished Library of Living Philosophers (Open Court, 1986).

Quine's eagerness to blur distinctions—boundaries, he once said, are for deans and librarians—underlies his second most famous controversy, also touched lightly in his autobiography. "To be is to be the value of a variable" is one of his most quoted remarks. Roughly, this means that if a formal system interprets X as a triangle, then the triangle (or any other abstract object) is as "real" (in a sense) as a watermelon. This tendency toward Platonic realism also brought Quine into sharp conflict with Carnap. Although Quine does not consider himself a Platonist, he insists that calling him a nominalist (one who thinks universals are merely words) is not right, either; indeed, he calls this one of the many misconceptions that have bedeviled him over the decades.

The Time of My Life abounds with amusing anecdotes about eminent philosophers who became Quine's friends, but none ever took the place of Ed Haskell, or Head Rascal, as Quine's two-year-old son once called him. Ed glides in and out of Quine's life like a curious shadow. The two met when they were undergraduates at Oberlin. Hitchhiking with his violin, Ed had earlier been picked up by a wealthy elderly woman who found him so engaging that she arranged to send him $100 a month for life. If this sum grew with inflation, it explains why Ed seems never to have had a job.

Quine describes his lifelong confidant as "ambitious, opinionated, contentious in the classroom, and rather shunned as an eccentric. . . . " In the thirties, Ed became an ardent Communist. A few years later, while a graduate student at the University of Chicago, Ed became attached to Count Alfred Korzybski's cult of general semantics. Quine had as little use then for the count as he had for communism. Soon, Ed was a strong anti-Communist, pinning his faith on a "unified science" that he hoped would save the world. Quine was trying once more to "apply the brakes" to Ed's "runaway ambition," when, incredibly, Ed became a booster of the Rev. Sun Moon! It was he who persuaded Moon to hold annual international conferences on unified science, at which Nobel prize winners were hornswoggled into speaking. Quine himself attended Moon's fourth conference, finding Moon funnier than any fundamentalist Bible thumper. At one point, Quine was about to admire physicist Eugene Wigner for getting up and walking out, but, no—Wigner was only going to the men's room.

Despite Quine's distrust of Ed's quirky enthusiasms, the two were constantly together for long walks and earnest talks. Quine calls Ed his "closest friend" and speaks of accepting speaking engagements only for the sake of their frequent reunions. Quine admits to weeping twice—with joy during his second marriage ceremony, with grief when he remembers how he fumbled a chance to join Ed in 1984 for a jaunt through western Texas. It was a project that Ed's failing health made impossible to renew.

Mitsumasa Anno

30

Imagine that you and a friend named Tom keep your eyes closed while a whimsical hatter puts a red hat on one of you and a white hat on the other. When you open your eyes and see a red hat on Tom, you know that your hat is white.

Suppose, now, there are *three* hats—two red, one white. Again the hatter puts a hat on you and Tom while your eyes and Tom's are closed. Eyes open, you see a red hat on Tom. After seeing your hat, Tom is certain *his* hat is red. What's the color of your hat?

The curious thing about this puzzle is that you have no way of knowing your hat's color until you hear Tom say his hat is red. You may not realize it, but if you solve the problem you have made use of elementary concepts in formal logic, combinatorial mathematics, and even binary arithmetic—which are absolutely fundamental to an understanding of modern mathematics and computer science.

Anno's Hat Tricks (Philomel, 1985), written by a Japanese mathematics professor and illustrated by the world-renowned artist Mitsumasa Anno, opens with the two puzzles just described. It then moves leisurely through a series of progressively more difficult problems involving three or more hats and a girl named Hannah who joins you and Tom. The text is so simple and crystal clear that any child who reads will be able to work on the puzzles. Mr. Anno had the happy thought of representing you, the reader, as Shadow-child, showing you in each picture as a shadow so that you cannot tell the color of your hat until you solve each problem.

If you and your children, or children you know, have not yet discovered Mitsumasa Anno, you are in for fantastic treats. For two decades this Tokyo artist, with his delightful style, his puckish hu-

This review originally appeared in the *New York Times Book Review,* 10 November 1985, and is reprinted here with permission. © 1985 by the New York Times Company.

mor, and his deep love of science and mathematics, has been creating absolutely marvelous books for children. I have been an admirer of his for a long time; indeed, I wrote the introduction to a book of his five years ago. Mr. Anno is sometimes likened to Maurits Escher because of his fondness for geometrical structures and visual mind-benders. In the book that introduced him to American readers in the 1970s, 'Anno's Alphabet,' each letter is drawn as an "impossible" object made of wood. Many of his other books swarm with optical tricks, upside-down pictures, hidden animals, mazes, mirror pictures, and endless other jokes and surprises.

Anno's Journey was the first of his classic "journey books." These are picture books without words, but every picture is so filled with wondrous details that both children and adults return to its pages, each time discovering amusing and beautiful things never before noticed. A child of any age can spend weeks studying the objects in *Anno's Flea Market*—his "love song to the past," as a reviewer described it last year in the *New York Times Book Review*—without exhausting its subtle nuances.

Many of Mr. Anno's dozens of books are designed to teach mathematics to very young children. Most of them are available, alas, only in their original Japanese, but we are fortunate to have some in English. Is there any better way to introduce children to the first twelve numbers than by giving them *Anno's Counting Book?* Its exciting pages depict scenes from January through December, each landscape filled with sets of objects to be counted by the number of the month. There is even a church clock in each picture that shows the time from 1 to 12. Addition and subtraction? There is no more pleasant way for a child to learn the meaning of these operations than by turning the pages of *Anno's Counting House.* Through die-cut windows they follow ten little people as they move themselves and their belongings from a furnished house to an empty house next door.

Anno's Mysterious Multiplying Jar is an incredible blue-and-white porcelain jar containing many things—islands and mountains and houses with cupboards that hold small jars exactly the original. Not knowing they are being taught, child readers learn what few adults know—the meaning of "factorial." Factorial 10, symbolized by an exclamation mark as $10!$, means $1 \times 2 \times 3 \times 4 \times 5 \times 6 \times 7 \times 8 \times 9 \times 10$. The product is 3,628,800. Mr. Anno's book shows how rapidly factorials grow and how they answer such questions as: How many ways can five soldiers stand in a row? (Answer: $5! = 120$.)

I know of no more painless way to introduce a bright child

to the meaning of what logicians call the "binary connective" of "if . . . , then" than to give the child *Anno's Hat Tricks*. You too will learn some elementary logic, including a way of diagramming deductive reasoning with a "binary tree," from the "Note to Parents and Other Older Readers" at the back of the book.

Did you solve the hat trick in our second paragraph? If so, try your reasoning skill on the puzzle that closes Mr. Anno's book. There are five hats—three red, two white. Tom says he does not know the color of his hat, but Hannah is sure hers is red. What's the color of your hat?

"This is a very hard question," Mr. Nozaki warns us. "If you can work the answer out by yourself, you're terrific. Please give it a try!"

31

WAP, SAP, PAP, and FAP

It has been observed that cosmologists are often wrong but seldom uncertain, and the authors of this long, fascinating, exasperating book, *The Anthropic Cosmological Principle* (Oxford University Press, 1986), are no exceptions. They are John Barrow, astronomer at the University of Sussex, and Frank Tipler, Tulane University mathematical physicist. Physicist John Wheeler provides an enthusiastic foreword. No one can plow through this well-written, painstakingly researched tome without absorbing vast chunks of information about QM (quantum mechanics), the latest cosmic models, and the history of philosophical views that bear on the book's main arguments.

Just what is this "anthropic principle" that has become so fashionable among a minority of cosmologists and is arousing such passionate controversy? As the authors make clear in their introduction, there is not one principle but four. Each is more speculative than the previous one, with the fourth blasting the authors out of science altogether into clouds of metaphysics and fantasy.

The simplest of the four is called (the authors are fond of acronyms) WAP, the *Weak Anthropic Principle*. Although it goes back to Protagoras's famous declaration that "man is the measure of all things," its modern cosmological form seems first to have been stated by the physicist Robert Dicke in the late 1950s. As Barrow and Tipler readily admit, it is a trivial tautology, totally noncontroversial. It merely proclaims that because we exist the universe must be so constructed as to allow us to have evolved. The laws of nature clearly must be such as to permit, if not actually force, the formation of CHON (carbon, hydrogen, oxygen, and nitrogen), the four elements essential to life as we know it.

Does this mean that all life must be carbon based? Although the authors believe this, it does not follow from WAP. Even if there is noncarbon life elsewhere in the universe, the fact that *we* are carbon imposes a variety of tight restraints on the universe and its past. For example, the cosmos has to be about fifteen billion years old. Why? Because, the authors argue, elements necessary to organic molecules are cooked inside stars. If the universe were much younger, those elements would not be available and we would not be here. If the universe were much older, all the suns would have burned out, and we would not be here either.

WAP was invoked over and over again in earlier centuries by proponents of the design argument for God. It was WAPish to point out that if the earth were slightly closer to the sun, like Venus, water would boil away and carbon life would be impossible. If the earth were slightly farther from the sun, water would freeze and Earth would have the barren deserts of Mars. Theists liked to note that when water freezes it expands and floats on water and that, otherwise, lakes and rivers would freeze to the bottom in winter and all their life be destroyed. If Earth did not have an ozone atmosphere, animals could not survive ultraviolet radiation. And so on. Hundreds of similar arguments, most of them analyzed by Barrow and Tipler, seem to show that our universe, and especially our planet, were carefully designed to permit us to exist.

The close ties between WAP and the creation hypothesis impel the authors to write almost one hundred pages on traditional proofs of God from design. It is an excellent history, followed by almost as long a section on more recent teleological arguments. There are informative discussions of such post-Darwinian "process" thinkers as Henri Bergson, Samuel Alexander, Alfred Whitehead, and Charles Hartshorne, who see the universe as rolling toward a predetermined goal, as well as of "process theologians," who anchor the goal in God.

If WAP were all there is to the anthropic principle, the book would not have been worth writing. The authors continually stress the triviality of asserting no more than that the universe has a structure that makes carbon life possible. It is easy to caricature such retrograde reasoning. Instead of saying I am here because my parents met, I say that because I am here I know my parents met. How lucky for vacationers that sandy beaches are so near the sea! From the fact that I wear spectacles I can deduce the positions of my ears and nose. If a chess game ends with no queens on the board, I can infer with iron logic that both queens were captured. From the present state of the world one can obviously make all sorts of highly probable, sometimes certain, conjectures about its distant past.

But there is more to the anthropic principle than WAP. The next step is SAP, the *Strong Anthropic Principle*. Proposed in 1974 by the British cosmologist Brandon Carter, it maintains that life of any sort is impossible unless the basic laws of nature are exactly what they are. Consider gravity. If it were slightly stronger, the cosmos would long ago have stopped expanding, gone the other way, and collapsed into a black hole before galaxies could form. If gravity were slightly weaker, the cosmos would have expanded too rapidly to allow matter to clump into stars. In either case, you and I would not be here.

The strength of gravity is one of a dozen or more constants called dimensionless because they are independent of any measuring system. If one banana is twice as long as another, the number two is the same whether you measure the banana in inches or centimeters. It turns out that these fundamental constants are so finely tuned that if they varied ever so slightly, there could not be any carbon atoms and we would not be here. Instead of saying we are here because the constants are precisely what they are, SAP turns it around. We are here; therefore the constants had to be what they are.

In long chapters on physics, astrophysics, and biochemistry, often dense with technical details and mathematical formulas, Barrow and Tipler defend this reverse way of reasoning. A recurring theme is that SAP puts such narrow constraints on the constants and natural laws that it can lead to falsifiable predictions. Opponents of SAP take a dim view of this claim. The physicist Heinz Pagels, in a slashing attack on anthropic arguments in his article "A Cozy Cosmology,"[1] dismisses WAP and SAP as pure flimflam. Although they may occasionally suggest testable conjectures, they do so in such obvious ways that nothing is gained by elevating them into new principles.

According to Barrow and Tipler, the first successful anthropic prediction was made by the University of Chicago geologist Thomas Chamberlain. Geological evidence indicates a great age for the solar system. If the sun did not feed on atomic energy, Chamberlain guessed, it would have long ago burned out and we would not be

1. *The Sciences* (March/April 1985).

here. Chamberlain guessed right, but did he do anything except apply ordinary reasoning?

Writing just before the Barrow and Tipler book was published, Pagels cites a more recent example. England's Stephen Hawking and Barry Collins once invoked the anthropic principle to explain why the universe is so isotropic—the same in all directions. If it were less so, matter would not condense into galaxies and we would not be here. This, says Pagels, explains nothing. By contrast, the new inflationary models of the Big Bang hypothesis of the origins of the universe actually do provide a plausible mechanism for isotropy. In old Big Bang models the initial explosion would have produced permanent irregularities. In the inflationary models, immediately after the bang the universe jumps from a trillionth the size of a proton to about the size of a softball. This sudden inflation smoothes out all irregularities, leaving an isotropic cosmos expanding at its present slow rate. In light of such speculations, the anthropic principle seems irrelevant. Surprisingly, Barrow and Tipler agree. They are strongly critical of Hawking and Collins for what they see as a misuse of the principle.

Similar efforts to use SAP as a tool for investigating the constants have been equally feeble, Pagels continues. Meanwhile, the new unified-field theories really are providing significant explanations of why the constants are what they are. WAP and SAP are so needless that they raise a new mystery. "How can such a sterile idea," Pagels asks, "reproduce itself so prolifically?" He suspects it may be because scientists are reluctant to make a leap of faith and say: "The reason the universe seems tailor-made for our existence is that it *was* tailor-made. . . . Faced with questions that do not neatly fit into the framework of science, they are loath to resort to religious explanations; yet their curiosity will not let them leave matters unaddressed. Hence, the anthropic principle. It is the closest that some atheists can get to God."

If one leaves aside the hypothesis of a transcendent Creator or of a Mind that either is the universe or permeates the universe, what alternatives are left? Barrow and Tipler consider several possibilities.

One is the startling view that only one kind of universe is possible—the one we know. This was skillfully defended by the Harvard chemist Lawrence Henderson in two books that were largely ignored until recently: *Fitness of the Environment* (1913, reprinted in 1970 by Harvard University Press) and *The Order of Nature* (1917). Leibniz argued exactly the opposite. He believed an infinity of universes are logically possible and that God selected the one he liked best.[2]

The authors discuss several variations of the many-possible-worlds view. Other universes could have the same laws as ours but entirely different histories depending on different initial conditions before the Big Bang. Parallel worlds could flourish side by side in our familiar three-dimensional space or in higher spaces, but because of limitations on the speed of light, no contacts between them are possible even if they are all in our space. We need not, however, assume infinite space. Alternate worlds could follow one another in some sort of supertime. Each explodes into existence, expands, contracts, and vanishes in the Big Crunch to be followed (whatever that means) by another fireball.

John Wheeler has a stupendous vision in which an infinity of universes pop in and out of existence, each with a randomly determined set of laws. Every logically possible universe appears an infinite number of times. (If an infinity of bridge hands are dealt, every possible distribution of the cards will be dealt an infinite number of times.) Of course only a tiny subset of these possible worlds will have forces and particles that permit life. This naturally emasculates any argument from design to God. It is not surprising we are in a universe that allowed us to evolve. How could it be otherwise?

The wildest of all variants of the infinite-universes theme, designed to counter the standard Copenhagen interpretation of QM (named for the city where Niels Bohr worked), is the many-worlds interpretation. In the Copenhagen view, the central mystery is what happens when a quantum system is measured. Take the case of a single particle. Every particle has associated with it a set of probability waves in an artificially constructed multidimensional space. A single expression, called the wave function, gives the probabilities that a particle will assume each of its possible states when it is measured. Before measurement, all possible states of the particle are said to be mixed in some sort of weird potential sense. Not until the particle is measured does nature "decide," by pure chance, what value to give a variable. At that instant the wave function is said to "collapse" from an indefinite to a definite state.

2. Isaac Newton held the same opinion. Here is a famous passage from his *Opticks:* "It may also be allowed that God is able to create particles of matter of several sizes and figures, and in several proportions to space, and perhaps of different densities and forces, and thereby to vary the laws of nature, and make worlds of several sorts in several parts of the universe. At least, I see nothing of contradiction in all this."

This notion of wave-function collapse leads to all sorts of paradoxes, of which the EPR paradox (after the initials of Einstein and two associates) has recently become the most notorious. A particle and its antiparticle can be simultaneously created by an interaction that sends them in opposite directions. Regardless of how far apart they get, perhaps light years in distance, they remain "correlated." If, for instance, one particle is measured for the direction of its spin, the wave function for the pair instantly collapses and the other particle acquires an opposite spin. Since neither particle has a definite spin until one is measured, and since there is no known causal connection between the pair, how does the other particle "know" what spin to acquire?

Einstein was deeply troubled by what he called the spooky telepathic aspect of this famous thought experiment, which he believed showed that QM is not a complete theory. Niels Bohr strongly disagreed. Most quantum experts still side with Bohr, though a growing number are beginning to suspect that the old maestro may have been right after all. The EPR paradox has recently been confirmed by several laboratory tests that could not have been made in Einstein's day. Perhaps this has awakened physicists to a fuller awareness of the paradox's deep implications.

The many-worlds interpretation dissolves the mystery of the EPR and similar paradoxes by denying that wave functions ever collapse. For this simplification, however, a horrendous price is paid. At every instant when a collapse seems to occur, the entire universe is said to split into parallel worlds, each containing one of the possible outcomes of measurement. At every instant billions upon billions of such splits take place. There is no communication between these worlds. We cannot tell that we are constantly splitting into duplicate selves because our consciousness rides smoothly along only one path in the endlessly forking chains. This splitting process is completely deterministic, perhaps guided by one monstrous wave function that keeps expanding but never collapses unless there is a God outside the universe to observe it.

The many-worlds interpretation has been called a beautiful theory that nobody can believe. Nevertheless, a number of eminent physicists, including Wheeler and Hawking, have taken it seriously, at least as a way of interpreting QM that removes its thorniest difficulties. Although Wheeler has withdrawn his support of the theory, Barrow and Tipler are defenders. "The wave function collapse postulated by the Copenhagen Interpretation is dynamically ridiculous," they write, "and this interpretation is difficult if not impossible to apply in quantum cosmology. We suggest that the Many-

Worlds Interpretation may well eventually replace the Statistical and Copenhagen Interpretations just as the Copernican system replaced the Ptolemaic. . . . Physicists who think in terms of the Copenhagen Interpretation may become handicapped in thinking about quantum cosmology." Poor old Bohr! Too bad he did not think of how to solve the problems of measurement by letting the universe copy him billions of times!

We come now to Wheeler's radical version of SAP, which the authors call PAP, or the *Participatory Anthropic Principle*. No universe can exist in a strong sense, Wheeler maintains, unless it contains conscious observers. This view rests on the fact that when a wave function collapses, the measuring instrument (a device or a person) becomes part of a larger system. All the potential states remain mixed as before until the larger system is measured.

Erwin Schrödinger, who disliked QM even though he helped get it started, invented a famous cat paradox to ridicule wave-function collapse. The cat is in a closed box with a mechanism that will kill the cat when it is triggered by a quantum event such as a click in a Geiger counter. The click has an equal probability of occurring or not within, say, an hour. At the end of the hour QM seems to say that until the cat is "measured" by someone looking into the box, the cat is neither alive nor dead. The two states remain mixed until an observer collapses the cat's wave function.

The paradox gets worse when you realize that even when an observer looks into the box he at once becomes part of a still larger system in which the cat's two states continue to be mixed until someone observes the observer. This is called the "paradox of Wigner's friend" after Eugene Wigner, a physicist who is troubled by it. It obviously leads to an infinite regress of observers. Wigner avoids the regress by cutting it whenever a chain of events is registered in a conscious mind. This raises more difficulties. Is the cat conscious enough to end the regress if it is not killed? If the mechanism merely chops off a leg? Although the chain ends for Wigner's friend when he opens the box, it does not end for Wigner until he observes his friend, so the regress does not really go away.

These solipsistic speculations have led Wheeler to the view that our universe is a participatory one in which reality is a collusion between minds and whatever is out there, perhaps only a bare mathematical field. For Wheeler the universe does not exist except in a pale mathematical sense unless it contains conscious observers. Here again the continuum of minds in the animal world raises disturbing problems. Einstein said he could not believe the moon's reality depends on being observed by a mouse. If a mouse will do, why not a bee?

Wheeler's view seems to be that a universe becomes real only when it is structured at the Big Bang so that it eventually can observe itself through conscious minds. Or perhaps reality can be thought of as a spectrum. The universe grows more real as life evolves to higher forms. In either case, Wheeler's vision sees conscious life as essential if a universe is to be more than a mathematical abstraction. His vision is close to Bishop Berkeley's "to be is to be perceived," except that Wheeler, unlike the Irish cleric, does not restore the external world by having it observed by God. Not for a moment did Berkeley, as sometimes said, doubt the external world's full reality. He only denied it was material. Indeed, he argued anthropically. Because we and the external world surely exist, there must be a God.

Barrow and Tipler move on from PAP to what they call FAP, the *Final Anthropic Principle*, but not before a long attempt to show that ETIs (Extraterrestrial Intelligences) do not exist. For many years Tipler has been arguing strenuously in both technical and popular articles that there are compelling grounds to assume that life on a level above microorganisms exists nowhere else in our galaxy, perhaps nowhere else in the universe. This has understandably brought him into sharp conflict with Frank Drake, Philip Morrison, Carl Sagan, and other scientists who strongly support SETI (the Search for Extraterrestrial Intelligence). Tipler is convinced that this search is a foolish waste of money.

Those who agree usually base their reasoning on the fact that long sequences of improbable events appear to be required for life even to get started. First there must be a sun with a planet on which conditions are extremely close to those on Earth. Even if it is assumed that such planets exist, the chance probability that a self-replicating molecule will arise on one of them may be vanishingly small. Finally, even if such a molecule does arise spontaneously, another sequence of improbable events must occur if it is to evolve into anything as intelligent as a fish or a bird.

In *Science Year* (1973) Wheeler had a science-fiction story called "Beyond the Black Hole." In it a character called Audrey, with whose views I assume Wheeler identifies, comes to this conclusion:

> Let's carry what you are saying to the logical extreme, Fred. It takes a very narrow squeak for a cycle of the universe to permit life at all, even at one place. If life had originated in more than one place, that

would have meant that the universe was larger and longer-lived than necessary. The creation of life would be more "expensive" than it needed to be. So the chances are overwhelming that Earth is the sole outpost of life in the universe, and we had no right to expect to find life on Zeta Zeta. Am I wrong in my reasoning?

To such reasoning Tipler has added a curious new argument that goes like this. There are planets in the cosmos millions of years older than Earth. If there is intelligent life on any of them, its technology would be far more advanced than ours. We know from experience that there is an overwhelming desire to explore the universe and that this is possible. It can best be done, Tipler claims, by what he calls von Neumann machines after the mathematician John von Neumann, who first proved that self-replicating robots can be constructed. Superbeings on other planets would surely build such robots and give them an intelligence equal to or surpassing their own. These robots would multiply at an explosive rate. In a short time they would be poking their spaceships into every corner of the galaxy. Because we see no signs of them (the authors have no interest in UFOs), ETIs do not exist.

Sagan is understandably infuriated by this reasoning. "Absence of evidence is not evidence of absence," he and William Newman said in their paper on "The Solipsist Approach to Extraterrestrial Intelligence."[3] In any case, the only way we can know is by searching. "We have an alternative denied to the medieval scholastics; we are able to experiment."

Why has the notion that we are alone in the galaxy been gaining ground? Partly, I suspect, because of the shock of finding no traces of life on Mars, partly because of a revival of theism that seems to be taking place among intellectuals. Whatever the reasons, those who have shared Tipler's unbelief in ETI include many distinguished evolutionists (Alfred Russel Wallace wrote an entire book about it) and such top physicists as Enrico Fermi and Freeman Dyson. "I find that the universe in some sense," wrote Dyson in his autobiography, "must have known that we were coming."

In his collection of essays, *The Flamingo's Smile,* Stephen Jay Gould attacks what he calls the "moth-eaten" arguments of the anthropicists. He accuses Tipler of misinterpreting what evolutionists mean when they speak of improbable events. They mean only that

3. *Quarterly Journal of the Royal Astronomical Society,* (Vol. 24, 1983), pp. 113–21. Reprinted in *Extraterrestrials,* edited by Edward Regis, Jr. (Cambridge University Press, 1985).

it is highly improbable evolution would take precisely the paths it has taken on Earth. They do not mean that once life starts the steps would not lead to intelligence. No one expects to find animals on another planet that duplicate beasts on Earth, but there are no good reasons for assuming that evolution could not take many paths to other forms of intelligence.

The assumed absence of ETI leads Barrow and Tipler to FAP, their *Final Anthropic Principle*. Although life probably exists only on Earth, now that it has begun FAP says it will be impossible to destroy. Otherwise, the universe would lose all its observers—and by PAP would have demolished itself! In the author's FAP fantasy, life is now taking its first faltering steps toward colonizing the universe. This is likely to be completed by intelligent von Neumann machines. The authors expect about half the universe to be colonized by the time our universe reaches the limit of its expansion and starts the other way. The red shift of stars turns to blue. Colonization goes on until the entire cosmos teems with computer life.

If there is an infinity of other universes, presumably these events will take place in all of them that permit life. Borrowing from Teilhard de Chardin, the Catholic paleontologist, the authors posit an Omega Point that will be the end of Everything. Here are the book's final sentences: "At the instant the Omega Point is reached, life will have gained control of *all* matter and forces not only in a single universe, but in all universes whose existence is logically possible; life will have spread into *all* spatial regions in all universes which could logically exist, and will have stored an infinite amount of information, including *all* bits of knowledge which it is logically possible to know. [123] And this is the end."

Footnote 123 is: "A modern-day theologian might wish to say that the totality of life at the Omega Point is omnipotent, omnipresent, and omniscient!" The God of Moses, Jesus, and Mohammed will finally have come into being. Instead of creating all the universes, however, it is the other way around. The universes got together and created the Almighty. This places the authors within the tradition of Samuel Alexander who, in his masterwork *Space, Time, and Deity* (1920), put forth the notion of a finite God who is slowly developing and growing in perfection as the universe evolves. Their eschatology is even closer to that of several famous science-fiction yarns. In Isaac Asimov's "The Last Question" a supercomputer evolves in hyperspace into a deity who creates a new universe to replace the old one that wore out and died the "heat death" dictated by thermodynamics.

What should one make of this quartet of WAP, SAP, PAP, and

FAP? In my not so humble opinion I think the last principle is best called CRAP, the Completely Ridiculous Anthropic Principle.

Postscript

Frank Tipler, in a long letter to *The New York Review of Books* (December 4, 1986), accused me (among other things) of misunderstanding the many-world's interpretation of QM. I replied in two words: "I'm speechless!" For a discussion of how the many-world's interpretation relates to traveling backward in time and for some comments on Tipler's plan for a possible time machine, see the first chapter of my *Time Travel and Other Mathematical Bewilderments* (W. H. Freeman, 1987).

"Newton, forgive me," Einstein wrote in an autobiographical essay. "You found the only way which, in your age, was just about possible for a man of highest thought and creative power." What was Einstein asking forgiveness for? That is the subject of this splendid book, *Was Einstein Right?* (Basic Books, 1986), by Clifford Will, a physicist at Washington University in St. Louis.

The subject is general relativity, or Einstein's theory of gravity, and how it has repeatedly been confirmed since 1960 by major experiments. But first some background.

The simplest kind of relative motion was fully understood by the ancients. If you are on a large ship that moves at a steady rate through calm waters, you can toss a ball back and forth as easily as on shore, even though the ball follows complicated paths relative to the stationary land. Of course the land is not really stationary. The earth rotates and goes around the sun. The sun moves relative to the stars of our Milky Way galaxy. The galaxy in turn rotates and moves relative to other galaxies. Is there some sort of fixed reference frame against which a final, absolute motion can be defined?

Yes, said Newton. Motion is absolute with respect to space. Before Einstein, physicists trying to explain how light can go through a vacuum—waves seem to require a medium to transmit them—postulated a fixed substance called the ether. Experiments had shown that the speed of light through this imagined ether was independent of the speed of its source. It should be possible, therefore, to determine the absolute motion of the earth with respect to a "stagnant" ether by measuring the speed of light in different directions on the earth's surface. The famous Michelson-Morley ex-

Reprinted with permission from *The New York Review of Books*. © 1986 Nyrev, Inc. This review originally appeared 4 December 1986.

periment of 1887 proved this could not be done. There was no trace of an "ether wind" generated by the earth's motion.

In 1905, apparently unaware of the Michelson-Morley results, Einstein published his special theory of relativity. Essentially, it discarded the notion of an ether and asserted that light (or any other portion of the electromagnetic spectrum) has a constant relative velocity regardless of the motion of an observer. If you travel alongside a light beam at half the speed of light, or even go the opposite way, the beam will always go past you at about 186,000 miles per second. Granting this absolute value for the speed of light relative to "the observer"—an observer moving in any direction at any speed—all sorts of strange effects involving space, time, mass, and energy—including the famous formula $E = mc^2$—inexorably follow.

The special theory concerned only motions in one direction at a constant speed . What about accelerated motions, such as the violent inertial effects astronauts undergo when their ship blasts off or the inertia that caused a young earth to bulge at its equator? Inertia is the tendency of bodies to stay at rest or continue moving in a straight line unless an external force acts on them. It is hard to walk on a merry-go-round because inertia acts as a centrifugal force that propels you outward. When the rotating earth was forming, the stronger centrifugal force near its equator, where matter moved faster than near the poles, gave the earth its present oblate shape. Do not these effects establish absolute motion? If you rotate a bucket of water, said Newton, inertia causes the water's surface to become concave. Is this not proof that the bucket, not the world, is rotating?

No, said Einstein in his general theory of relativity, published in 1915. There is no way to distinguish between a rotating bucket and a motionless bucket with a universe whirling around it. Only the relative motion of bucket and universe is "real." We say the bucket rotates because it is much simpler to take the universe as fixed, just as it is simpler to say I stand on the earth instead of saying the earth rests on the bottom of my shoes. We choose the Copernican system over the Ptolemaic not because it is true and the other false but because it is enormously simpler.

Generalizing the special theory to all motion was a far greater creative leap than the special theory. Had Einstein not published his paper on the special theory, others would soon have reached the same conclusions. Indeed, Henri Poincaré in France and H. A. Lorentz in the Netherlands almost got there ahead of Einstein. But the general theory was such an amazing jump of the imagination, into totally unexplored territory, that physicists are still in awe over how Einstein managed it.

At the heart of the general theory is what Einstein called the principle of equivalence. It asserts that gravity and inertia are one and the same. If we take the universe as fixed, we say inertia caused the earth to bulge. If we take the earth as fixed, the rotating universe generates a gravity field that caused the bulge. The relative rotation of earth and universe creates a single force field that can be called gravitational or inertial depending on our choice of a reference frame. Had someone suggested to Newton that inertia and gravity were two names for the same force, he would have thought that person crazy.

The principle of equivalence made it necessary, as Professor Will adroitly explains, to replace Newton's "flat" three-dimensional Euclidian space with a non-Euclidian space of four dimensions. The fourth coordinate is time, and the curvature of space-time varies from place to place. Gravity ceases to be a "force" in the Newtonian sense. The earth goes around the sun not because the sun tugs it but because the sun warps space-time in such a way that the earth finds an elliptical orbit the simplest, "straightest" path it can take in space as it hurtles ahead in time. As John Wheeler likes to say, the stars tell space-time how to bend, and the bends tell the stars and other objects where to go.

In general relativity this distortion of space-time propagates like a wave, traveling at the speed of light. Quantum mechanics requires that gravity waves have their associated particles called gravitons. A variety of weird events occurring outside our galaxy, all carefully covered by Professor Will, strongly imply the existence of gravity waves. However, gravity is such a weak interaction that no one has yet detected its waves in a laboratory. Claims to have done so remain unreplicated. More sensitive tests are now under way, and it would be hard to find a physicist who doubts that gravity waves and gravitons eventually will be detected.

When a relativist says it is permissible to deem Newton's bucket stationary and the universe spinning, what does he mean by "universe"? Does he mean no more than the totality of stars and other celestial objects, or does he include a space-time structure, a metric field, that would be there even if the material universe disappeared? If the universe contained nothing but Newton's bucket, could the bucket rotate? If so, would its water experience inertia?

Mach's principle, named by Einstein for the nineteenth-century Austrian physicist and philosopher Ernst Mach, maintains that if the bucket were all there is, it would be meaningless to say it rotates. From this point of view inertia arises because there is accelerated motion (rotation is a form of acceleration) relative to the galaxies and other forms of matter and energy in the universe. (In Newton's day both Leibniz and Bishop George Berkeley had similarly argued against Newton that space is no more than a relation between bodies, with no reality by itself.) Although Mach lived to reject both special relativity and the existence of atoms, Einstein was greatly influenced by him and in his younger years was strongly attracted to the simplicity of Mach's principle. Later he became doubtful.

General relativity, Will makes clear, is compatible with both Mach's principle and the view that inertia arises wholly or in part from accelerated motion with respect to a metric field of space-time that is independent of the matter and energy it contains. Recent tests have tended to go against Mach's principle. Many pages of the book under review are devoted to a fantastic experiment designed by three Stanford physicists for an earth-orbiting laboratory. Based on the precessions of sophisticated gyroscopes, it could give a conclusive answer to the profound questions raised by Leibniz, Berkeley, and Mach. Planning for this test has been going on for more than two decades.

In 1962 when my *Relativity for the Million* was published—it was written for high school students—I said that, although the special theory was so completely vindicated it had become part of classical physics, evidence for the general theory remained feeble. Professor Will recalls an occasion that same year when a famous astronomer at the California Institute of Technology advised a graduate student to avoid relativity because it "had so little connection with the rest of physics and astronomy." Kip Thorne, the student, wisely ignored this advice. He is now at the forefront of research in the fast-expanding field called relativistic astronomy.

When I revised my book in 1976 for an edition retitled *The Relativity Explosion,* new tests of general relativity had been proliferating for fifteen years. Since 1976 more and better tests have been made. If you want to know details about these ingenious experiments and how the general theory has passed them all with what the author calls "flying colors," there is no better book available, none more clearly written for laymen or more up-to-date, than *Was Einstein Right?*

Einstein himself was supremely confident about his general theory because of its elegance and simplicity. Simplicity? Its complicated mathematics gave rise to endless cartoons, jokes, and anecdotes. The book recalls a story often told about Sir Arthur Stanley

Eddington, among the first of eminent British astronomers to accept general relativity. A colleague said to Eddington, "You must be one of three persons in the world who understands general relativity." Eddington was silent. "Don't be modest," said the colleague. "On the contrary," Eddington is said to have replied, "I am trying to think who the third person is."

In the light of observational and experimental results and the unification of gravity and inertia, the general theory is amazingly and beautifully simple. Professor Will recalls Einstein's joking remark that if tests ever decided against the theory it would only prove God made a mistake when he designed the universe. Of course Einstein knew that elegance is not enough to make a theory fertile. Early in the game he himself had proposed three ways of testing the basic ideas of general relativity. How much does light from distant stars bend when it passes close to the sun? Does the elliptical orbit of Mercury rotate on the plane at a rate which agrees with relativity? And is the wavelength of light shifted toward the red side of the spectrum when influenced by gravity?

Before 1960 all three tests gave only weak confirmations. Repeated attempts to measure the bending of starlight, as it grazed the sun during a total eclipse, were marred by huge margins of error. Measurements did confirm bending, but the degree of bend was impossible to pin down. Even Newtonian physics, Will reminds us, predicts the bending of light by gravity, though at only half the amount required by relativity. Mercury's orbit seemed to support Einstein, but again other explanations could not be ruled out. The gravitational red shift of light had almost no empirical support.

In the 1960s, Will writes, physicists suddenly found themselves in possession of fantastically powerful new tools. Atomic clocks of various kinds made possible incredibly accurate measurements of time. Laser instruments were perfected. Larger radio and X-ray telescopes were built. Faster computers made it easier to analyze complex data. Radar and laser light could be bounced off mirrors on the moon and off planets and satellites. What Will calls a renaissance of interest in general relativity soon emerged. At first the solar system was the new testing "laboratory." In the 1970s the laboratory enlarged to regions far beyond our galaxy.

Professor Will makes an important distinction between the basic ideas of general relativity, which physicists now take for granted, and the ten tensor equations Einstein finally provided as a way of measuring the curvature of space-time. If by "general relativity" we mean those equations, then in the 1960s many rival theories, with slightly different equations, were proposed. The most important was a theory devised by Princeton's Robert Dicke and his former graduate student Carl Brans. The Brans-Dicke theory, as it was later called, accepted all the central ideas of general relativity but modified Einstein's field equations by adding a second field. As a consequence, it made predictions that differed slightly from Einstein's.

Measurements of the sun's shape seemed to show that the sun was fatter at its equator than had been suspected, perhaps because its core rotated faster than its surface. When this oblateness was taken into account, the Brans-Dicke theory predicted the rotation of Mercury's orbit better than did Einstein's. In a chapter called "The Rise and Fall of the Brans-Dicke Theory" the author explains why knowledge of the sun's precise shape remains cloudy. The sun's brightness and the fact that it constantly throbs like a beating heart make its shape extremely difficult to determine. Some observations reported in 1985 seem to show that the sun's core rotates *more slowly* than its surface. In any case, support for the Brans-Dicke theory has been rapidly eroding.

The most precise measurements supporting Einstein over Brans-Dicke are described in the chapter "Do the Earth and the Moon Fall the Same?" Einstein's field equations require an absolute equivalence in the way all matter is influenced by gravity. "If we were to drop the Earth and a ball of aluminum in the gravitational field of some distant body," Will writes, "the two would fall at the same rate." A 1969 experiment, using lasers, verified that the earth and moon fall toward the sun with the same acceleration, and to a precision of one part in a hundred billion. Because the Brans-Dicke theory does not accept what is called the "strong equivalence principle," this test counted heavily against it. Had Einstein been told of its result, Will surmises, he would have replied, "Of course!"

Ephraim Fischbach of Purdue University has announced (too recently to be in Will's book) that he and his associates have found evidence for a hitherto undetected repulsive force which they call "hypercharge." If it exists, it would be much weaker than gravity—but could cause gravity to act differently on different kinds of matter. A feather would not fall in a vacuum with exactly the same acceleration as an iron ball. Such a new force would be a revolutionary challenge to the strong equivalence principle. Although Fischbach's claims have been widely publicized, most physicists remain skeptical.

Since 1960 numerous tests of the bending of light by gravity, as well as tests of the gravitational red shift, have strongly favored Einstein's equations. Will gives a detailed account of the first good measurement (in 1960) of this shift. The difference in shifting between the top of Harvard's Jefferson Tower and its base, where earth's gravity is stronger, confirmed Einstein's equations with a 10 percent margin of error. Later, the experiment was improved to an error of 1 percent. Measurements of the sun's influence on starlight were abandoned in the 1970s because results were too muddied by the sun's corona and other annoying factors. Different and more accurate tests have since been made in other ways, all in agreement with Einstein's field equations.

The famous twin paradox of relativity, involved in many science-fiction stories, is closely related to the gravitational red shift. It says that if one twin makes a long journey into space and returns, he will be younger than his brother who stayed home. If he goes far and fast enough, he could come back to find that centuries on earth had sped by. Time travel into the past remains logically flawed (if you went back to your childhood and shot yourself, you would simultaneously be alive and dead), but traveling to the earth's distant future is theoretically possible.

In the general theory of relativity the difference in aging can be explained by the fact that the stay-at-home twin does not move much relative to the universe whereas the traveling twin does. A handful of stubborn skeptics have argued in the past that relativity does not imply the twin paradox—or that if it does it must be wrong; but in the light of recent tests, their voices are seldom heard today. The book gives colorful details about how the twin paradox was validated in 1971 by flying two atomic clocks around the earth, one westward and the other eastward, then comparing them with an atomic clock that remained on the ground.

A fourth kind of test, not proposed by Einstein, involves the way gravity delays a light signal. Professor Will explains it with a rubber-sheet model. Put a heavy ball on the center of a flat elastic sheet supported at its perimeter. The ball will produce a depression—a three-dimensional distortion of the sheet's two-dimensional space. This causes a marble, placed anywhere outside the depression, to roll toward the ball. The ball does not pull the marble. The marble moves because of the sheet's curvature. If you imagine a light ray on the sheet, entering and leaving the depression, it will travel farther than it would if the sheet were flat. This is similar to what happens when light goes through a region strongly warped by a star's mass. Because the path has lengthened, there is what is called the Shapiro time delay, after Irwin Shapiro, who worked out the mathematics in the early 1960s. Complex measurements of this delay by Viking spacecraft have confirmed Einstein's field equations with an error of one part in one thousand. Will calls it "still the most accurate test of the theory ever performed."

In general relativity the strength of gravity never alters. However, the discovery that the universe originated in a monstrous explosion and has been expanding ever since raised the possibility that perhaps gravity is slowly weakening. This is especially plausible if Mach's principle holds. Paul Dirac, the British physicist who introduced special relativity into quantum mechanics, was among the first to suggest that gravity is getting weaker. The Brans-Dicke theory makes the same claim. A chapter titled "Is the Gravitational Constant Constant?" skillfully summarizes the latest experimental evidence that gravity is indeed constant, although definitive tests have yet to be made.

In brief, the book answers the question posed by its title with a resounding yes. Einstein *was* right. Not only have his equations been confirmed over and over again, but the general theory has become indispensable for understanding the incredible new objects that modern telescopes have detected: the pulsars believed to be fast-spinning neutron stars and the far-distant quasars suspected of having black holes at their centers because there seems no other way to account for their enormous energy output. The day has long passed, writes Will, when cosmologists can remain ignorant of relativity. Every year astrophysicists find new phenomena that only the general theory can explain. The most recent are the powerful gravity fields outside our galaxy that act like mammoth lenses, magnifying and refracting what is seen through them. Such lenses were predicted by Einstein in 1936.

Galileo and Newton made experiments, but the extraordinary thing about Einstein is that he made no experiments. Moreover, he was often unaware of significant tests that had strong bearings on his speculations. He just sat alone, thinking deeply about the secrets of the Old One, as he liked to call the universe. Newton was a devout Anglican who spent half his life struggling to unravel the mysteries of biblical prophecy. Einstein had no interest in any religion except in the sense that Spinoza, whose secular pantheism he admired, was religious. Yet he and Newton, in addition to their giant intellects and creative intuitions, shared a strong sense of wonder toward the Old One and of humility before the unanswerable riddle of existence. Both were Platonists in their conviction that what science knows is an infinitesimal portion of what it does not know.

Newton, in an often quoted passage, likened himself to a boy playing on the shore of a vast "ocean of truth," amusing himself by picking up a smooth pebble or a patterned shell. Einstein made the same point with a different metaphor. He told an interviewer that he thought of himself as a child who has entered an enormous library, its books written in many languages. He takes down one volume and manages to translate a few pages. What a far cry from those now trying to persuade us that physics is on the brink of discovering Everything!

Marvin Minsky's Theory of Mind

Marvin Minsky, a computer-science professor at MIT and cofounder of the institute's Artificial Intelligence Laboratory, is one of the towering architects of artificial intelligence (AI). His previous books have dealt with technical aspects of computer science and robotics. Now for the first time he has summarized in one volume (*The Society of Mind,* Simon and Schuster, 1986) his conjectures about how the natural brain, sitting silently inside its skull, has managed to evolve into such a marvelous tool for solving the problems humans face in their interactions with one another and with the outside world.

Most people imagine they have a soul or self, a tiny little creature inside their head who watches moving pictures carried to mental screens by sensory inputs. Would not such an elf, Minsky asks, require a smaller elf inside its head to make sense of the pictures— and so on into an unthinkable infinite regress?

The mind's hardware is, of course, the billions of neurons linked in a tangled net of fantastic complexity. Nobody knows how it works. Nevertheless, lack of such knowledge need not prevent trying to understand what it does. In Minsky's metaphorical theory, the mind is pictured as a society of billions of tiny agents, each totally mindless. "What we call a mind," wrote David Hume, "is nothing but a heap or collection of different perceptions." Minsky buys the heap but not the "nothing but." At every higher level of organization of agents, unexpected new skills emerge. The separate parts of an airplane cannot fly. The plane can. A single person cannot build a skyscraper. A construction crew can.

"The Society of Mind" is popularly written in a lucid staccato prose, sparkling with jokes and apt quotations, and put together in a way that reflects both the frames of a floppy disk and Minsky's social theory of the brain. Each page is a complete unit, joined in a web to other pages so that the book's rich insights emerge in nonlinear fashion as you read. At the back is a glossary of almost one hundred terms, most of them unfamiliar because Minsky made them up.

Elaborate terminologies have been set forth before by investigators of the mind, but Minsky's has an advantage over earlier ones. In recent decades, there has been a continuing upswing of experimental research on how humans, especially children, think and—more recently, on the results of AI. Building on these data, Minsky found it necessary to create a new vocabulary. He is aware, of course, that AI is in its babyhood, that today's robots are like windup toys, that only time will tell how useful his vocabulary will be.

Here are a few of its terms. An *A-brain* is a part of the brain joined to the external world by sensory organs. A *B-brain* is a part joined only to an A-brain. A *frame* is any mental representation with a set of terminals to which other structures can be attached. A *uniframe* is a frame that captures what a set of frames have in common, such as your concept of a bird. Uniframes are joined to *exceptions*. We all know, Minsky reminds us, that birds can fly—except for penguins, ostriches, dead birds, caged birds, and so on. Without the *exception principle,* uniframes would be useless.

Indeed, it is the fuzziness of concepts that makes it so difficult for machines to translate languages or to recognize patterns. "The plumber filled his pipe" is one of Minsky's amusing examples of verbal ambiguity. A baby quickly learns to know its mother, first by her smell, then by her face; but it is not easy to teach a machine to recognize a letter when it is given different shapes. Douglas Hofstadter somewhere recalls saying to Stanford's computer scientist Donald Knuth that AI's central problem is comprehending the essence of the letter *A.* Knuth responded: "And what is the letter *I?*"

K-line ("K" for knowledge) is a basic Minsky term. K-lines are wirelike agents that activate memories by joining agents to agents. (Coleridge called them hooks and eyes.) Modeling K-lines as strings allows Minsky to splatter his pages with topological diagrams that are sometimes trees, to represent *hierarchies,* sometimes graphs with loops, to represent *heterarchies.* Other colorful terms include the many varieties of *nemes* and *nomes,* demons that crouch in wait before they jump out to warn you of something, *difference-engines* that are goal-oriented agents, and *time blinking*—finding a difference between two mental states by activating them in rapid succession, like riffling a deck of cards to spot secret markings.

Memory is not a single entity. Names and faces are remembered by different processes, and short-term memory is not the same as long-term. Nor is intelligence one thing. It is a society of many diverse, imperfect skills. "What magical trick makes us intelligent?" Minsky asks, then answers: "The trick is that there is no trick." Learning, creativity, and genius are other vague omnibus words. Intuition holds no mystery. It is what we experience when we reach a conclusion, often wrong, after a sequence of unconscious reasoning steps. Consciousness reduces to "little more than menu lists that flash, from time to time, on mental screen displays," enabling you to recall a portion of your recent past. Philosophy's notorious mind-body problem similarly dissolves, as it dissolved long ago for so many philosophers and psychologists. Mind is what the brain does.

Is formal logic more complex than ordinary thinking? For Minsky, the reverse is true. It is easy to build machines that solve problems in formal systems, such as proving theorems in logic or mathematics, playing chess, and so on. It is enormously harder to teach a machine to build a house of blocks, pick up a cup of tea, or tie a shoelace. Common-sense thinking is "an immense society of hard-earned practical ideas," based mainly on analogies. In contrast, formal reasoning is both simpler and more easily misguided. Most snarks are green, every boojum is a snark, therefore most boojums are green. Wrong! Boojums, Minsky discloses, are albino snarks.

The brain, Minsky once said, is a computer made of meat. A lifelong science-fiction buff, he has no difficulty believing that far-future robots will talk and behave in ways indistinguishable from humans. They will develop consciousness, feel emotions, and produce good paintings, poetry, and music. Meat people may even achieve near-immortality by transferring their minds to robots. "There is not the slightest reason to doubt that brains are anything other than machines with enormous numbers of parts that work in perfect accord with physical laws."

From this perspective, free will is another illusion. Mental events are either determined by prior causes or random like the throw of a die. Neither is what we desperately desire free will to be. Alas, there is no third alternative.

Here I must, with great reluctance, part with Minsky's always stimulating speculations. I am among those willing to posit third alternatives even when they are rationally opaque. Minsky is puzzled by the brain's dark mysteries, but from my Platonic perspective he is not puzzled enough. Lest you suppose this an out-of-date idiosyncrasy, let me invoke a modern thinker whose linguistic theories have had considerable influence on Minsky and his distinguished associates. Noam Chomsky was asked about free will in a 1983 interview. Here are parts of his reply:

> People have been trying to solve the problem of free will for thousands of years, and they've made zero progress. They don't even have bad ideas about how to answer the question. My hunch . . . is that the answer to the riddle of free will lies in the domain of potential science that the human mind can never master because of the limitations of its genetic structure . . .
>
> It could well turn out that free will is one maze that we humans will never solve. We may be like the rat that simply is not designed to solve a certain type of maze. . . . Look, in principle there are almost certainly true scientific theories that our genetically determined brain structures will prevent us from ever understanding. . . . I'm not sure that I want free will to be understood.

"A year spent in AI," so runs an epigram by Yale's computer scientist Alan Perlis, "is enough to make one believe in God."

Postscript

The notion that free will is a transcendent mystery—that the "problem" of will is unsolvable—is out of fashion among today's thinkers, though it was commonplace in earlier epochs, both inside and outside traditional religious faiths. Kant, a philosophical theist, defended it on the grounds that the true self is what he called "noumenal," beyond our "phenomenal" time and space. A chapter on free will in my *Whys of a Philosophical Scrivener* marshalls the essential arguments.

It is not often today than nontheists such as Noam Chomsky adopt this view. Another recent and surprising example is Thomas Nagel, a philosopher at New York University. The following passages are from his chapter "Freedom," in *The View from Nowhere* (Oxford, 1986):

> I change my mind about the problem of free will every time I think about it, and therefore cannot offer any view with even moderate confidence; but my present opinion is that nothing that might be a solution has yet been described. This is not a case where there are several possible candidate solutions and we don't know which is correct. It is a case where nothing believable has (to my knowledge)

been proposed by anyone in the extensive public discussion of the subject.

The difficulty, as I shall try to explain, is that while we can easily evoke disturbing effects by taking up an external view of our own actions and the actions of others, it is impossible to give a coherent account of the internal view of action which is under threat. When we try to explain what we believe which seems to be undermined by a conception of actions as events in the world—determined or not— we end up with something that is either incomprehensible or clearly inadequate.

This naturally suggests that the threat is unreal, and that an account of freedom can be given which is compatible with the objective view, and perhaps even with determinism. But I believe this is not the case. All such accounts fail to allay the feeling that, looked at from far enough outside, agents are helpless and not responsible. Compatibilist accounts of freedom tend to be even less plausible than libertarian ones. Nor is it possible simply to dissolve our unanalyzed sense of autonomy and responsibility. It is something we can't get rid of, either in relation to ourselves or in relation to others. We are apparently condemned to want something impossible.

Nature's patterns are marvelous mixtures of order and chaos. The moon looks perfectly round, but through a telescope, its circumference is jagged. In quiet air, the smoke from a pipe rises in a fairly straight plume, then quickly dissolves into disheveled swirls. The earth follows an almost perfect ellipse around the sun, but its coastlines, rivers, and lightning bolts twist anywhither.

Now there is nothing new about these observations, but about twenty-five years ago, mathematicians and physicists began to make some remarkable discoveries. Physicists found that many natural phenomena that display random behavior are not immune to simple mathematical theorems. And mathematicians found that trivial equations can describe the movement of a point that is indistinguishable from random behavior. There is order behind the chaos of the physical world, and chaos contaminates elementary algebra. Called chaos theory, it is now one of the hottest new areas of research. Its unexpected laws are finding important applications to such diverse, seemingly random phenomena as weather prediction, air and liquid turbulence, oscillating chemical reactions, heart disorders, cancer growth, epilepsy, evolution, computer breakdowns and the great red spot on Jupiter. *Chaos: Making a New Science* (Viking, 1987), by the *New York Times* science writer James Gleick, is the first popularly written book about this fascinating, rapidly growing discipline. It is a splendid introduction. Not only does it explain accurately and skillfully the fundamentals of chaos theory, but it also sketches the theory's colorful history, with entertaining anecdotes about its pioneers and provocative asides about the philosophy of science and mathematics.

The book opens with the story of how Edward Lorentz, a meteorologist at MIT, discovered in the early 1960s the first of what came to be called "strange attractors." But first, what is an ordinary attractor? Imagine a rigid pendulum, swinging back and forth without friction or air resistance. If you plot its perpetual movement on a coordinate plane (one axis for position, the other for time), the bob's trajectory is a circle. This circle is the bob's attractor, the path to which its behavior is confined. If you add friction to the pendulum, it soon stops. Its path now graphs as a spiral. The attractor is the fixed point at the spiral's center.

What Lorentz discovered was a completely new kind of attractor. Imagine a wheel on which buckets hang. They fill with water at a steady rate, and lose water at a steady rate through a small hole. Lorentz proved that when the two rates have certain values, the wheel's motion becomes utterly chaotic—and in a way closely related to convection currents. If the wheel's motion is plotted as the movement of a point in what physicists call a three-dimensional phase space, the point's trajectory—the wheel's strange attractor—turns out to be an infinitely long double spiral that resembles butterfly wings. The curve never self-intersects. This means that the wheel's behavior, as it randomly alters speed and direction, never repeats in a predictable way. Lorentz, Gleick tells us, actually built a wheel model in his basement to convince skeptics that such a simple deterministic system would behave chaotically.

When a physical system is described by linear equations—those with no exponents higher than one—the equations graph as straight lines and the system's behavior is orderly and predictable. Equations with exponents greater than one are called nonlinear. Most natural phenomena are described by nonlinear equations, especially by the nonlinear differential equations of calculus. It is only in nonlinear systems that chaotic behavior appears, constrained or "bounded" by strange attractors when the behavior is modeled by points in a phase space. It turns out that these chaotic attractors are fractals—curves of infinite complexity but self-similar in the strange sense that, like coastlines, they display the same form when portions are magnified.

Many of these strange attractors arise from a process called "periodic doubling" or "bifurcation" that is similar to what happens when taffy is stretched and folded by a taffy machine or when an ordered deck of playing cards is repeatedly shuffled. Spots close together in the taffy (or cards in the deck) rapidly diverge and wander about in a random way. If certain nonlinear systems are modeled in phase space, an analogous stretch-and-fold process occurs as the system "bifurcates into chaos." A strange attractor called Smale's horseshoe (after mathematician Stephen Smale) was the earliest attractor related to this particular road (there are many others) to chaos. Bifurcation is controlled by a universal constant (an irrational number with a value of 4.6692 +) called a Feigenbaum number, after its discoverer, Mitchell Feigenbaum. Trajectories diverge so rapidly on this road to chaos that tiny variations of initial conditions are quickly magnified. Meteorologists call it the butterfly effect because it has been said that the flutter of a butterfly's wings could trigger a causal cascade that would produce a tornado. It is this effect that renders long-range weather prediction intrinsically impossible.

Lorentz's discovery sparked an intensive search for other chaotic systems bounded by other strange attractors. Because a computer screen provides a way of modeling chaos if the latter is simple enough, the computer quickly became such a powerful research tool that much of chaos theory has come from mathematicians sitting alone at computer consoles and experimenting with low-order nonlinear equations. When different colors are used to distinguish aspects of chaos, extraordinarily complex and beautiful patterns emerge. Gleick's book has eight color plates of these dazzling designs. Especially exquisite are the fractal Julia sets (after the French mathematician Gaston Julia) that separate regions of chaos from what are called basins of attraction. Named strange repellers because they repel points rather than attract, they are intimately related to the famous Mandelbrot set discovered in 1979 by IBM's Benoit Mandelbrot. It was Mandelbrot who coined the term fractal and who was the first to investigate fractals in depth.

The Mandelbrot set, generated by an absurdly simple procedure that involves the continual squaring and adding of complex numbers, differs from all other fractals in the following bizarre way. Each new magnification introduces unpredictable change. The set has been likened to an incredible jungle of exotic flora and fauna which mathematicians have only started to explore. Because its delicate filigree of flames and spirals goes to infinity, its properties may never become fully known. It is a thing of strange beauty and awesome complexity, containing endless depths of wild surprises. Gleick rightly regards it as the most mysterious object in geometry. "A devil's polymer," Mandelbrot has called it.

Ten years ago there was a great burst of interest in a new branch of topology known as catastrophe theory. Some have speculated that this enthusiasm was generated in part by the theory's colorful name. Is the term *chaos* playing a similar role? Would chaos theory have been embraced so fervidly by young computer hackers if it had been called, say, "nonlinear bounded randomness"?

Some chaos evangelists have suggested—and Gleick is inclined to agree—that chaos theory is a revolution destined to alter physics as drastically as did relativity theory and quantum mechanics. Or will chaos fade into just another—albeit fascinating aspect of probability theory or perhaps into an aspect of fractal theory? Only time will tell.

The world is colors and motion, feelings and thought . . . and what does math have to do with it? Not much, if "math" means being bored in high school, but in truth mathematics is the one universal science. Mathematics is the study of pure pattern and everything in the cosmos is a kind of pattern.

In the above quotation, the first paragraph of Rudy Rucker's *Mind Tools: The Five Levels of Mathematical Reality* (Houghton-Mifflin, 1987), observe the word *pure*. Mathematical patterns are pure, timeless concepts, uncontaminated by reality. Yet the outside world is so structured that these patterns in the mind apply to it with eerie accuracy. Nothing has more radically altered human history than this uncanny, to some inexplicable, interplay of pure math and the structure of whatever is "out there." The interplay is responsible for all science and technology.

Perhaps it is a dim awareness of the explosive role of mathematics in altering the world, together with the low quality of math teaching in this country, that accounts for the growing number of books intended to teach mathematics to those who hated it in school. The two books here under review are general surveys, in the tradition of such popular classics as Edward Kasner and James Newman's *Mathematics and the Imagination*. Unlike most such surveys, each book is organized around a unifying concept.

For Eli Maor, an Israeli mathematician now at Oakland University in Rochester, Michigan, the unifying concept of *To Infinity and Beyond: A Cultural History of the Infinite* (Birkhäuser, 1987) is infinity. *Finite mathematics,* a term that has come into recent use, is precalculus math in which infinity is avoided as much as possible; yet even in the most elementary math there is no way to escape completely from the concept. As Maor points out, counting numbers go on forever, and straight lines are endless in both directions. Textbooks on finite math have chapters on probability, but what is meant when you say the odds are equal that a flipped coin will fall heads or tails? "We tacitly assume," writes Maor, "that an infinite number of tosses would produce an equal outcome."

Maor begins his admirable survey with the concept of limit. In one of Zeno's notorious paradoxes, a runner can not get from A to B until he first goes half the distance. Now he must run half the remaining distance, then half the still remaining distance, and so on into an infinite regress. Because at any time the number of distances yet to be traversed is infinite, how can he reach B? Worse than that, how can he begin? If the distance is sixteen miles, he must first run eight miles. To go eight he must go four. Again, the halves form an infinite sequence. How does he get started? Of course mathematicians are no longer troubled by such paradoxes of motion, but it is impossible to resolve them without a clear notion of the limits of infinite sequences of magnitudes in both time and space.

Maor's well-chosen examples are wide-ranging. Archimedes determined the value of pi (the ratio of a circle's circumference to its diameter) by calculating the perimeters of inscribed and circumscribed regular polygons. By increasing the number of sides of these nested figures he was able to squeeze the value of pi between inside and outside polygons that came closer and closer to the limit of a circle. In this way he got pi correct for the first time to what today we call two decimal places. In 1986 a Japanese supercomputer calculated pi to more than 134 million digits.

At present no one knows whether certain patterns, say a run of a hundred sevens, occur somewhere in the nonrepeating endless decimals of pi. Are we entitled to say the run is either there (wherever "there" is) or not there? Here the concept of infinity generates a curious split in the philosophy of mathematics. A Platonic realist would answer "of course," the run of sevens is there or not, but there are mathematicians called constructivists who will have none of this. The either/or cannot be asserted, they insist, until such a run is actually found, or until someone proves in a finite number of steps that the run must or cannot "sleep" in pi, as William James once put it. It is, of course, legitimate for a constructivist to say that a run of a hundred sevens does or does not exist in the first billion decimals of pi, because there are algorithms (procedures) for answering this question in a finite number of steps.

Proving whether certain types of numbers belong to finite or infinite sets is a major ongoing task of number theory. Maor gives Euclid's elegant proof that the number of primes is infinite. (A prime is an integer greater than 1, divisible only by itself and 1.) Twin primes are primes that differ by two, such as 3,5 and 11,13.

Are they infinite as well? Twins of monstrous size have been found by computers, but whether there is an infinity of them remains unanswered. Maor reports that in 1982 a computer software company offered $25,000 for the first proof of this old conjecture.

The harmonic series, which has so many applications in physics, is $1/1 + 1/2 + 1/3 + 1/4 + \ldots$, where the ellipsis indicates an infinity of the reciprocals of the counting numbers. (The reciprocal of x is $1/x$.) As the number of terms increases, the partial sums become larger. Do these sums converge (approach a limit) or diverge (increase without limit)? Because the terms are increasingly smaller, one suspects the sum converges. Amazingly, it does not though the divergence becomes increasingly sluggish. It takes 12,367 terms to reach a sum that exceeds 10; to exceed 100 the number of terms required has forty-four digits.

Maor provides many curiosities involving this remarkable sequence. If you remove all the terms that contain a specified digit in the denominator, the series converges. For example, if you remove all fractions that contain 9, the sequence converges on a sum slightly less than 23. Suppose you remove all fractions with denominators that are not prime. The series still diverges. On the other hand, the reciprocals of twin primes (assuming they are infinite) have been shown to converge.

Maor's discussion of numbers comes to a climax with chapters on Georg Cantor's revolutionary discovery that it is possible to define "transfinite" numbers that stand for an infinite heirarchy of infinities. The smallest—Cantor called it aleph-null—counts the integers, as well as any infinite subset of the integers. For instance, there are as many primes as there are integers. The proof is simply to put the two sets of numbers into one-to-one correspondence:

$$1,2,3,4,5, \ldots$$
$$2,3,5,7,11, \ldots.$$

Any set of objects that can be put into correspondence with the integers is called countable. Cantor was able to show that the set of all integral fractions is countable but that the set of irrational numbers (numbers such as pi and the square root of 2) that cannot be expressed as integral fractions is not countable. Cantor called the number that counts the real numbers (rational and irrational) aleph-one, or C (for continuum), because, as Maor shows, it counts the number of points on a line segment. Cantor believed that 2 raised to the power of aleph-null is the same as C, and he proved that an endless ladder of alephs can be generated simply by raising 2 to the power of higher and higher alephs.

Turning to geometry, Maor covers a variety of fascinating topics, such as infinitely long "pathological curves" that enclose a finite area, and surfaces of infinite area that surround a finite volume. A section on inversion explains how a circle or sphere can be turned inside out to put all its points into correspondence with all outside points on an infinite plane or in infinite space. There is an old mathematical joke about how to catch a tiger. You invert the space outside an empty cage. This puts the tiger (along with everything else) into the cage.

A chapter on the Dutch artist Maurits Escher reproduces many of his pictures (some in color) that involve infinity, such as his marvelous mosaics of birds and animals that tile the infinite plane. (The plane is said to be tiled if the shapes completely cover it, without gaps or overlaps, like the hexagonal tiles of a bathroom floor.) The jacket of Maor's book has an Escher drawing of a globe covered with loxodromes. These are helical paths followed by ships and planes that travel at a constant angle (not a right angle) to the earth's meridians. The paths spiral around the poles, making an infinity of revolutions until they strangle the poles.

Maor ends his survey by leaving pure math for the disheveled outside world. Discussions of modern cosmology and particle physics raise deep questions about the infinitely large and the infinitely small. Does space-time extend forever, or is it finite but unbounded, as the surface of a sphere would be for flatlanders living on it? Are there other universes out there in some sort of hyper space-time? Does the infinitely small stop with a truly fundamental particle (the latest speculation is that the basic units are infinitesimal strings), or is matter an infinite regress of endlessly smaller entities, like an infinite nest of Oriental wooden dolls?

Rudy Rucker, who holds a doctorate in set theory, is a professor of computer science at San Jose State University in California. He is well known to science-fiction readers for his far-out fantasies, including *White Light,* a novel based on Cantor's alephs. Another novel assumes that as you shrink down into smaller and smaller levels of reality you eventually enter the same universe you started from. Rucker's previous nonfiction books, including *Infinity and the Mind* and *The Fourth Dimension,* mix mathematics with occasional bizarre science-fiction themes. *Mind Tools,* a survey of math organized around the modern concept of information, is a similar blend.

Rucker divides mathematics into what he calls five archetypes or modes of thought: Number, Space, Logic, Infinity, and Information. A section of his book is devoted to each mode, the first four approached from an information perspective. To explain the modes, Rucker considers a human hand.

From the perspective of number, the fingers model the integer 5, but scores of other numbers count such quantities as hairs, wrinkles, cells, lengths of fingers, areas of nails, weight, temperature, blood-flow rate, electrical conductivity, and so on. Viewed as space, the hand is a three-dimensional solid. Because it lacks holes, it is topologically equivalent to a ball. (Topology studies properties that remain the same when an object is continuously deformed.) The hand's blood vessels branch in a pattern that mathematicians call a tree. Parts of the hand's surface are concave, parts convex. From a logic point of view the hand is a machine about which all sorts of "if, . . . then" statements can be made: if it clenches, knuckles get white; if it touches fire, it jerks away; if it digs in dirt, fingernails get black.

Infinity enters when you consider the hand as an abstract solid with an uncountable infinity of points. As an actual solid, if the nested-dolls conjecture holds, it may have an infinity of components. Viewed as information, the hand grew in accord with detailed instructions coded by the body's DNA. Information about the hand's past is embodied in such traces as scars and freckles. How many questions would someone have to ask about your hand to build a replica? What is the shortest computer program that would give this information?

Rucker likes to substitute familiar words for technical jargon. Instead of saying the world is a mixture of discreteness and continuity, he speaks of spottiness and smoothness. The usual references to the wave/particle duality of quantum mechanics are replaced by talk of lumps and bumps. Sometimes an electron acts like a discrete lump, sometimes like a bump in the shifting patterns of a wave field. Which is more fundamental, a particle or its field? This, says Rucker, is like asking which is more fundamental, a person or society? He invokes Niels Bohr's famous aphorism: "A great truth is a statement whose opposite is also a great truth." Bohr called this the principle of complementarity. He was so intrigued by the Oriental yin-yang symbol of complementarity that he put it on his coat of arms.

Following in the mental steps of his great-great-grandfather, the German philosopher Hegel, Rucker is a monist who believes that in some ultimate sense, like the circle that surrounds the yin and yang, all is One. There is no need, he writes, to distinguish either a particle from its field or a person from society. "Reality is one, and language introduces impossible distinctions that need not be made." Need not be made? To a pluralist—William James for example—the distinctions *have* to be made, not just because language forces them but because that is how the universe is fragmented. I once ran across a couplet by some unknown poet whose name I long ago forgot, though not the lines:

> If all is One,
> Who will win?

In some transcendent sense, monism may prevail, but the white light of Hegel's Absolute is stained by Shelley's dome of many-colored glass, and without the distinctions we could not think, talk, or live. Indeed, Rucker could not have written his book without thousands of distinctions in pure mathematics. As for the outside world, nothing is perfectly smooth. Everything has lumps.

Such metaphysical animadversions need not hinder a pluralist from enjoying Rucker's lively explorations. His number section tells how to use your fingers as flip-flops for binary counting. This leads to a discussion of logorithms, figurate numbers (numbers modeled by spots in patterned arrays), giant numbers, and the numerology of interesting numbers from 1 to 100. Ninety-one is particularly interesting. It counts the spots in a triangular array of thirteen spots on the side, the spots in a hexagonal array of six spots on the side, and the number of balls in a pyramid with six on the side of its square base. It is the sum of the cubes of 3 and 4, and when you write it in base-9 notation it is 111. Twenty-three is the smallest integer Rucker found relatively boring.

The section on space allows Rucker to introduce tiling theory, with special attention to an extraordinary discovery in 1974 by the British mathematical physicist Roger Penrose. Penrose found a pair of quadrilateral figures, usually called "kites and darts" because of their shapes, that tile the plane in only a nonperiodic way. A periodic tiling is one on which you can outline a region that tiles the plane by translation (shifting without rotating or reflecting), in the manner of the bricks that tile a brick wall. On a nonperiodic tiling, no such region can be outlined. It is of course possible to tile the plane nonperiodically with replications of a single shape as simple as a triangle or square, but such shapes also tile periodically. Whether there exists a single shape that will tile *only* nonperiodically is one of the major unsolved problems of tiling theory.

The amazing thing about Penrose's kites and darts is that

the only way they will cover the plane, without gaps or overlaps, is nonperiodically. Mathematicians—notably John Conway, now at Princeton University—at once began finding all sorts of astonishing properties of Penrose tiling, when a few years ago a wholly unexpected event took place. Crystals were constructed with atoms arranged in a nonperiodic pattern based on a three-dimensional analog of Penrose tiles! Hundreds of papers have since appeared about these strange "quasicrystals." It is a superb instance of how a discovery in what can be called pure recreational mathematics suddenly found a totally unexpected application to the shaggy world "out there." [1]

The same Conway invented the most profound of all computer recreations, the cellular-automaton game of Life. A cellular automaton is a structure of cells, each of which can assume a certain number of states. At each "tick" of time, the states simultaneously alter according to "transition rules" that govern the passage of information to a cell from a specified set of "neighbors." Cellular-automata theory is now a hot topic on the fringes of math, with many applications to robot theory and artificial intelligence. Edward Fredkin, at MIT, has conjectured that the universe itself may be one vast cellular automaton. As Rucker points out, this vision is similar to Leibniz's dream of a cosmos composed of isolated monads that "have no windows," and are incessantly changing in obedience to transition rules decreed by God. Viewed this way, the universe is playing a computer game so awesomely complex that the fastest way anyone will ever be able to predict its future states is just to let the game go on and see what happens.[2]

Discussions of classical curves (including some with such splendid names as Pearls of Sluze and the Nephroid of Freeth) lead Rucker into the exciting new field of fractals, a remarkable kind of irregular pattern that Benoit Mandelbrot was the first to investigate in depth. A fractal is an infinitely long curve or infinitely complex pattern that always looks the same if you keep enlarging portions of it. Mandelbrot called them fractals because he found an ingenious way to assign them fractional space dimensions. During the last ten

1. You can find out more about Penrose tiling in chapter 10 of *Tilings and Patterns,* a beautiful book by Branko Grünbaum and G. C. Shephard (W. H. Freeman, 1986).

2. Even the ridiculously simple transition rules of Life, concerning cells with only two states and eight neighbors, create patterns impossible to predict. An entire book about Life and its philosophical implications is William Poundstone's *The Recursive Universe* (Morrow, 1984).

years, following Mandelbrot's brilliant leads, fractals have found hundreds of applications in science and aesthetics.[3] A coastline, the surfaces of mountains, and the surface of the moon are familiar approximations of fractals. As a camera gets closer to the moon, photographing smaller and smaller craters, the surface still looks the same. Computer programs are now generating fractal music and fantastic fractal landscapes for science-fiction films. The topic propels Rucker into one of his wild conjectures, but you will have to consult his chapter "Life is a Fractal in Hilbert Space" to get the details.

Rucker's next section, on logic, begins with Aristotle's syllogisms, followed by the propositional calculus and the predicate calculus, the two lowest levels of symbolic logic. Next comes a stimulating discussion of Kurt Gödel's famous undecidability proof that in any formal system complicated enough to include arithmetic, true theorems can be stated that can not be proved within the system. For instance, Goldbach's conjecture—that every even number is the sum of two primes—could, in the light of Gödel's theorem, be undecidable. If so, mathematicians may be doomed never either to find a counterexample or to prove the conjecture true.

Rucker examines Gödel's theorem from his five perspectives; he ties the discussion into the theory of Turing machines (idealized computers), and a theorem of Alonzo Church's that says that no algorithm (step-by-step procedure) exists that will in a finite time tell whether an arbitrary statement in a complex formal system (one more complex than the propositional calculus) is true. The section ends with musings on how dull life would be if Gödel's and Church's theorems did not hold. "Our world is endlessly more complicated than any finite program or any finite set of rules. You're free, and you're really alive, and there's no telling what you'll think of next."

The section on information carries Rucker into questions about infinity. Cantor's alephs are explained; then, going the other way, the infinitesimally small numbers of a modern approach to calculus called nonstandard analysis are explained. Bishop George Berkeley ridiculed the infinitesimal magnitudes in the calculus of Newton and Leibniz, but now, thanks to the labors of Abraham Robinson, infinitely small quantities are as respectable as Cantor's alephs. The section leads into subtle information theorems recently

3. On fractals, see Benoit Mandelbrot's masterpiece, *The Fractal Geometry of Nature* (W. H. Freeman, 1982), and *The Beauty of Fractals*, by H. O. Peitgen and P. H. Richter (Springer-Verlag, 1986).

established by Gregory Chaitin and his colleague Charles Bennett. Rucker paraphrases their theorems in a characteristically cryptic way:

> Speaking more loosely, Chaitin showed that we can't prove that the world has no simple explanation. Bennett showed that the world may indeed have a simple explanation, but that the world may be so logically deep that it takes an impossibly long time to turn the explanation into actual predictions about phenomena.
>
> To make it even simpler: Chaitin shows that we can't disprove the existence of a simple Secret of Life, but Bennett shows that, even if someone tells you the Secret of Life, turning it into usable knowledge may prove incredibly hard. The Secret of Life may not be worth knowing.

Hegel had a compulsion to group ideas into triads of thesis, antithesis, and synthesis. His great-great-great grandson's book ends not with a triad but a pentad:

> My purpose in writing *Mind Tools* has been to see what follows if one believes that everything is information. I have reached the following (debatable) conclusions.
>
> 1) The world can be resolved into digital bits, with each bit made of smaller bits.
> 2) These bits form a fractal pattern in fact-space.
> 3) The pattern behaves like a cellular automaton.
> 4) The pattern is inconceivably large in size and in dimensions.
> 5) Although the world started very simply, its computation is irreducibly complex.
>
> So what is reality, one more time? An incompressible computation by a fractal CA [cellular automaton] of inconceivable dimensions. And where is this huge computation taking place? Everywhere; it's what we're made of.

The Curious Mind of Allan Bloom

In the 1930s, when Robert Hutchins was president of the University of Chicago, he and Mortimer Adler raised a great ruckus about the decay of college education. In *The Higher Learning in America* and in later books, speeches, and articles, Hutchins pleaded for a return to liberal education grounded in familiarity with the great literature and philosophy of the past.

Hutchins lived to see all the trends he denounced with such vigor and wit steadily increase. Our universities, he said in 1954, have become "highclass flophouses where parents send their children to keep them off the labor market and out of their own hair." Now comes Allan David Bloom, a professor of social thought at the University of Chicago, to echo and amplify the Hutchins-Adler rhetoric in his *Closing of The American Mind* (Simon and Schuster, 1987). The book is subtitled "How Higher Education Has Failed Democracy and Impoverished The Souls of Today's Students." Bloom's former student, Saul Bellow, wrote the foreword. In spite of Bloom's scholarly style and erudition—and to everyone's amazement—his book was for many months on the *New York Times's* nonfiction bestseller list.

Professor Bloom is the author of a volume about Shakespeare's politics and is translator of two classics on education: Plato's *Republic* and Rousseau's *Emile*. His *Closing of the American Mind,* a powerful, idiosyncratic indictment of everything he finds wrong on today's campuses, has obviously struck a raw nerve, especially among parents who have seen their offspring graduate college with minds as empty as they were in high school.

Like Hutchins before him, Bloom blames the mediocrity of

This review originally appeared in *Education and Society,* Spring 1988, and is reprinted here by permission of *Education and Society* and the Anti-Defamation League of B'nai B'rith.

college teaching on a pervasive moral and philosophical relativism. In reaction to the certainties of past ages of faith, relativism abolishes all absolutes except the absolute of being free from absolutes. Bloom sees this great opening as a great closing. Because there is no way to define the good life, colleges no longer urge students to seek it. Instead of introducing young people to the wisdom of the past, colleges offer a flea market of unrelated courses from which students select whatever they find easiest and most to their liking.

Unfortunately, Bloom weakens his arguments by unrestrained caricature. He tells us that professors and students have lost sight of what the Constitution calls inalienable rights. He sees black power as an effort to obtain superior rights, not equality. Affirmative action promotes the racism our Founding Fathers tried to "defang." Government loans and quotas have flooded the universities with poorly prepared blacks whose teachers are afraid to give them low grades and whose presence is a leading cause of sagging educational standards. Although relativism has promoted an admirable belief in racial equality, there has been little social integration. Tables in eating halls still separate into black and white. Bloom attributes this in part to a sense of shame among black students for their special treatment—and to a smoldering white resentment.

Bloom is all for openness to other cultures, provided it stimulates a search for standards; but if there is no fundamental human nature, with needs common to all societies, the search becomes meaningless. Good versus evil gives way to an "I'm okay, you're okay" attitude that discourages interest in ethics and politics. Because there are no standards for the good life, no Platonic vision of an ideal society, the basis for the social activism of the sixties has evaporated. Students have turned inward, absorbed with self, concerned only with making money and enjoying movies, sex, and music.

Sex, however, has become casual, passionless, and "flat-souled." Romance has gone the way of binding contracts. Students have forgotten how to say "I love you" except when they dump a bed partner. Bloom is struck speechless when he sees a couple, who have been roommates throughout their college years, part with a handshake. Relationships have turned gray and amorphous. Students do not date; they live in groups with no more sex differentiation than "animal herds when not in heat." The desire for marriage has diminished—and along with it the motive for gallantry. "Why should a man risk his life to protect a karate champion?"

Women are rapidly gaining equality, as Plato thought they should; but Bloom thinks the feminists go too far in trying to obliterate natural distinctions. "Law may prescribe that the male nipples be made equal to the female ones," he declares in one of his most embarrassing aphorisms, "but they still will not give milk." The feminists are trapped between conflicting loyalties. They want sexual freedom, yet they resent being portrayed in girlie magazines as bimbos. They ridicule their mothers' advice—"He won't respect you or marry you if you give him what he wants too easily"—and then they wonder why they are losing both respect and marriage proposals.

Bloom is also down on the feminists for what he perceives as their indifference to the great books. Instead of reading them to learn how earlier ages coped with male chauvinism, they dismiss them as useless relics of a male-dominated past. To Bloom's annoyance, they have even persuaded Bible translators to replace all masculine pronouns for God with neuter pronouns, as if great books should be rewritten to avoid offending the latest sensibilities.[1]

Like all conservatives, most of whom have hailed his book as a modern classic, Bloom sees the monogamous family as essential to any good society and sees the rising divorce rate as "America's most urgent social problem." Students from broken homes are filled with suppressed "rage, doubt and fear." Many undergo long, ineffective mental therapies, financed by guilt-ridden parents. There are no hints in Bloom's book of the terrible quarrels that prevailed when divorce was difficult—or of the damage this inflicted on children. There is no hope that new freedoms may lead to happier homes.

Loss of aesthetic standards parallels the loss of truth and goodness. Students no longer can distinguish "the sublime from trash." This is most obvious, Bloom thinks, with respect to music. Except for a small elite, classical music is dead. There is no escape from rock's boom, thumpa, boom in dormitories, cars, on TV and movie screens, in concert halls, and blasting out of Walkmans. "As long as they have the Walkman on, they cannot hear what the great tradition has to say. And, after its prolonged use, when they take it off, they find they are deaf."

I would guess that Bloom's blistering attack on rock—rivaled only by a chapter in Jerry Falwell's *Listen, America!*—is a major

1. Bloom's perpetual, compulsive use of male pronouns reflects his opposition to the growing practice of eliminating sexist language from secular books. On the first two pages of his preface I counted seventeen male pronouns that refer to college teachers. Although his university's president is a woman, no female faculty members are visible in his book. In a seven-page chapter titled "The Self" I circled fifty-eight uses of "man" and "men" when bachelor Bloom really means humanity.

reason for his book's success. Word about it has spread and almost every reader will relish it. (The people who will not probably will not be reading the book.) What is the culmination of our vast technology? "A pubescent child whose body throbs with orgasmic rhythms, whose feelings are made articulate in hymns to the joys of onanism or the killing of parents; whose ambition is to win fame and wealth in imitating the drag-queen who makes the music."

Bloom's final chapter, a stirring defense of the "good old great books," is pure Hutchins except for a failure to mention how college athletics divert money and energy from everything a university is supposed to do. Hutchins is never mentioned. Mortimer Adler is cited only to praise his business acumen in promoting the Great Books set he edited. Indeed, Bloom calls the Hutchins-Adler Great Books movement an amateurish "cult," marred by a "coarse evangelistic tone."

Nevertheless, Bloom believes that great books should be the core of every liberal education. He can understand why scientists are indifferent to the science classics. Unlike the liberal arts, science is a cumulative enlargement and refinement of knowledge. Studying Newton's *Principia,* for example, will not teach a physics student anything he cannot learn more easily from a modern textbook. It is the hostility to literary and philosophical classics by liberal arts professors that puzzles Bloom, who maintains that in the humanities, now a "submerged old Atlantis," book reading has degenerated into an elevation of criticism over content. Bloom sees the latest French fad, deconstructionism, as the ultimate neglect of what great works of art and literature say in favor of how they should be examined. He predicts that deconstructionism will soon deconstruct here as it already has in Paris.

Bloom is as coy as was Hutchins about revealing his own metaphysical posits. There are constant references to the "soul," even to the "perfect soul," and on page 137 he distinguishes the soul from both body and mind. Kant is praised for viewing the soul as an utterly mysterious entity that "stands outside the grasp of science." Does Bloom think of the soul as Kant and Plato did (and as Aristotle did not)—that is, as a personality that survives the body?

Bloom clearly believes there are objective moral standards; but how to recognize them remains vague. Is he an emotive ethicist who rests morality on nothing firmer than human desires, or does he think reason and science can support a naturalistic ethics that cuts across all cultural barriers? Must we look to God as the ground of morals? "Real religion and knowledge of the Bible have diminished to the vanishing point," Bloom complains; but what on earth does he mean by "real religion"? Atheists, he insists, have a "better grasp of religion" than those sociologists who are fascinated with the "sacred." Bloom likens the latter to a man who keeps a "toothless old circus lion around the house . . . to experience the thrills of the jungle." Does the lion with teeth exist? If so, Bloom is silent on where to look for it.[2]

As for knowledge of the Bible fading, imagine Bloom in debate with Falwell or Jimmy Swaggart about what the Good Book says. He would quickly discover that today's fundamentalist leaders know the Scriptures as thoroughly as did Saint Augustine and Martin Luther. Nowhere does Bloom consider either the baleful effect of fundamentalism on American education or the equally debilitating influence of the occult revolution. About half our college students, polls show, believe in Satan, angels, and astrology.

Bloom devotes a chapter to arguing that today's cultural relativism springs from the popularization of German philosophy, especially the teachings of Nietzsche and such of his heirs as Freud, Max Weber, and Martin Heidegger. (Nietzsche has a longer list of references in Bloom's index than any other person.) "The self-understanding of hippies, yippies, yuppies, panthers, prelates and presidents has unconsciously been formed by German thought of a half-century earlier." Nietzsche has had more influence on the American left, Bloom actually believes, than Karl Marx, now obsolete and boring. He sees German nihilism behind radical violence, even in the popularity of the song "Mack the Knife." There are horror tales of Bloom's experiences with student violence at Cornell— and strong condemnation of the cowardice of Cornell officials and professors. Terrorism around the world and the thirst for bloody revolution are both consequences of Nietzsche's evil ideas.

This is Teutonic baloney. Nietzsche was something of a rage among U.S. intellectuals when H. L. Mencken wrote a book about

---

2. Bloom's respect for the Bible is mystifying. I was unable to learn anything about his religious upbringing, but on page 60 Bloom has nothing but praise for the Bible's influence on his grandparents. "Their home was spiritually rich because all the things done in it . . . found their origin in the Bible's commandments . . . and the commentaries on them. . . . I am not saying anything so trite as that life is fuller when people have myths to live by. I mean rather that a life based on the Book is closer to the truth, that it provides the material for deeper research in and access to the real nature of things." On pages 374–75 he deplores the way colleges teach the Bible only as literature. "To include it in the humanities is already a blasphemy, a denial of its own claims." Such remarks make sense coming from a conservative Catholic or Protestant. Why Bloom feels compelled to make them is almost as unfathomable as Mortimer Adler's lifelong reverence for Thomas Aquinas.

his philosophy in 1908, but his influence on native relativism even then was minimal. The stronger influence came from sociologists and anthropologists, who may have been impressed by German metaphysics but whose relativism flowed mainly from their investigations. Books such as William Sumner's *Folkways* were more influential than any book by Nietzsche.

Nowhere does Bloom seem aware that American philosophy has for half a century been tramping to the beat of British skepticism and empiricism. Hume has had far more effect on American philosophy than have the German metaphysicians. For instance, our greatest thinkers—Charles Peirce, William James, and John Dewey—regarded Hegel as the source of everything wrong in philosophy. (Incidentally, the only influential twentieth-century philosopher or theologian mentioned in the book is Dewey.) Protestant, Catholic, and Jewish theologians have also reacted violently to what they consider the Hegelian sin of hubris. Bloom's pommeling of German metaphysics reads as if it had been written in 1916, the year Santayana published *Egotism and German Philosophy.*

Bloom has a knack for writing wisely and eloquently, then suddenly uncorking something foolish. I will cite two Bloomers. On page 52 he sees Descartes and Pascal as opposites, representing the eternal conflict between reason and revelation. But both men were great creative mathematicians, and is there any exercise of reason purer than mathematics? As for revelation, both were devout Catholics. The main difference: Descartes thought unaided reason could prove such things as God's existence. Pascal, the better thinker, was sure it could not.

On page 106 we read: "To strangers from another planet, what would be the most striking thing is that sexual passion [among our youth] no longer includes the illusion of eternity." Does Bloom suppose that on other worlds there are humanoids with sex organs and marriage rituals like ours?

I closed *Closing* with unbounded admiration for the clarity of Bloom's style and for his quixotic courage in battling educators who will ignore arguments that are fundamentally sound. I also had a strong feeling of *deja vu,* because I had heard it all before. Hutchins and Adler fought the same fight when I was an undergraduate at the University of Chicago. At least Hutchins achieved one of his goals. He got rid of the football team.